T0331151

Sparse Graphical Modeling for High Dimensional Data

This book provides a general framework for learning sparse graphical models with conditional independence tests. It includes complete treatments for Gaussian, Poisson, multinomial, and mixed data; unified treatments for covariate adjustments, data integration, network comparison, missing data, and heterogeneous data; efficient methods for joint estimation of multiple graphical models; and effective methods of high-dimensional variable selection and high-dimensional inference. The methods possess an embarrassingly parallel structure in performing conditional independence tests, and the computation can be significantly accelerated by running in parallel on a multi-core computer or a parallel architecture. This book is intended to serve researchers and scientists interested in high-dimensional statistics and graduate students in broad data science disciplines.

Key Features:

- A general framework for learning sparse graphical models with conditional independence tests
- Complete treatments for different types of data, Gaussian, Poisson, multinomial, and mixed data
- Unified treatments for data integration, network comparison, and covariate adjustment
- Unified treatments for missing data and heterogeneous data
- Efficient methods for joint estimation of multiple graphical models
- Effective methods of high-dimensional variable selection
- Effective methods of high-dimensional inference

Monographs on Statistics and Applied Probability

Series Editors:
F. Bunea, R. Henderson, N. Keiding, L. Levina, N. Meinshausen, R. Smith

Recently Published Titles

Sufficient Dimension Reduction
Methods and Applications with R
Bing Li

Large Covariance and Autocovariance Matrices
Arup Bose and Monika Bhattacharjee

The Statistical Analysis of Multivariate Failure Time Data: A Marginal Modeling Approach
Ross L. Prentice and Shanshan Zhao

Dynamic Treatment Regimes
Statistical Methods for Precision Medicine
Anastasios A. Tsiatis, Marie Davidian, Shannon T. Holloway, and Eric B. Laber

Sequential Change Detection and Hypothesis Testing
General Non-i.i.d. Stochastic Models and Asymptotically Optimal Rules
Alexander Tartakovsky

Introduction to Time Series Modeling
Genshiro Kitigawa

Replication and Evidence Factors in Observational Studies
Paul R. Rosenbaum

Introduction to High-Dimensional Statistics, Second Edition
Christophe Giraud

Object Oriented Data Analysis
J.S. Marron and Ian L. Dryden

Martingale Methods in Statistics
Yoichi Nishiyama

The Energy of Data and Distance Correlation
Gabor J. Szekely and Maria L. Rizzo

Sparse Graphical Modeling for High-Dimensional Data
Faming Liang and Bochao Jia

For more information about this series please visit:
https://www.crcpress.com/Chapman–HallCRC-Monographs-on-Statistics–Applied-Probability/book-series/CHMONSTAAPP

Sparse Graphical Modeling for High Dimensional Data

A Paradigm of Conditional Independence Tests

Faming Liang and Bochao Jia

CRC Press
Taylor & Francis Group
Boca Raton London New York

CRC Press is an imprint of the
Taylor & Francis Group, an **informa** business

A CHAPMAN & HALL BOOK

Designed cover image: © Faming Liang and Bochao Jia

First edition published 2023
by CRC Press
6000 Broken Sound Parkway NW, Suite 300, Boca Raton, FL 33487-2742

and by CRC Press
4 Park Square, Milton Park, Abingdon, Oxon, OX14 4RN

CRC Press is an imprint of Taylor & Francis Group, LLC

© 2023 Taylor & Francis Group, LLC

ISBN: 9780367183738 (hbk)
ISBN: 9781032481470 (pbk)
ISBN: 9780429061189 (ebk)

DOI: 10.1201/9780429061189

Typeset in Nimbus
by codeMantra

Publisher's note: This book has been prepared from camera-ready copy provided by the authors.

To our families

Contents

Preface

Coming with the new century, the integration of computer technology into science and daily life has enabled scientists to collect a great amount of high-dimensional data, such as omics data and portfolio data, for which the sample size is typically much smaller than the dimension. Sparse graphical modeling is at the heart of high-dimensional data analysis, which allows one to break the high-dimensional system into simpler parts for further statistical inference. Motivated by this insight, a variety of approaches have been proposed for sparse graphical modeling, such as node-wise regression [118], graphical Lasso [6, 49, 193], Bayesian regularization [54], and many of their variants, which, in general, enforce the model sparsity via regularization.

A fundamental concept to sparse graphical modeling is conditional independence, which rationalizes the sparsity of graphical models. Motivated by this observation, Liang et al. [93] proposed an equivalent measure of partial correlation coefficients for a large set of Gaussian random variables, based on which the high-dimensional Gaussian graphical model can be constructed (via a multiple hypothesis test) as based on the true partial correlation coefficients. The embarrassingly parallel structure of this approach in calculating the equivalent partial correlation coefficients makes it highly scalable. Later, Liang and coauthors extended this multiple conditional independent tests-based approach to learning graphical models for Poisson data [74], mixed data [183], and non-Gaussian data [101]. They also extended the approach to jointly learning multiple graphical models for the data collected under distinct conditions [77]. With the recent statistical development for high-dimensional missing data problems [92], they further extended the approach to learning sparse graphical models with missing data [92] and heterogeneous data [73]. As a consequent application of sparse graphical modeling, they also developed a so-called Markov neighborhood regression method for high-dimensional statistical inference with the aid of sparse graphical models [94, 160]. In particular, with the aid of the sparse graphical model of the explanatory variables, the Markov neighborhood regression method has successfully broken the high-dimensional inference problem into a series of low-dimensional inference problems where conventional statistical inference methods, such as those on p-value evaluation and confidence interval construction, can still be applied. Markov neighborhood regression has brought a new insight to high-dimensional inference problems.

This book is to provide a systematic and up-to-date treatment for high-dimensional graphical models under the framework of multiple conditional independent tests. Chapter 1 provides a short introduction to high-dimensional graphical models and introduces a unified paradigm of conditional independence tests for high-dimensional statistical inference as well. Chapters 2–6 provide a systematic description for the multiple conditional independent tests-based approaches and their applications for estimation of graphical models under various scenarios, including the cases with standard Gaussian data, missing data, heterogeneous data, Poisson data, and mixed data. Chapter 7 describes how to extend the approach to joint estimation of multiple graphical models. Chapter 8 describes how to extend the approach to estimation of nonlinear and non-Gaussian graphical models. As a consequent application of sparse graphical modeling, Chapter 9 describes how to conduct statistical inference for high-dimensional regression with the aid of sparse graphical models.

This book is intended to serve three audiences: researchers specializing in high-dimensional graphical modeling, scientists interested in data science, and graduate students in statistics, biostatistics, computational biology, or broad data science disciplines. This book can also be as a textbook for a special topic graduate course with the focus on sparse graphical modeling.

Faming Liang
West Lafayette, Indiana
November, 2022

Authors

Dr. Faming Liang is a Distinguished Professor of Statistics at Purdue University, West Lafayette, Indiana, USA. He earned his PhD degree in statistics from the Chinese University of Hong Kong, in 1997, and has been recognized as a fellow of the American Statistical Association (ASA), the Institute of Mathematical Statistics (IMS), and an elected member of the International Statistical Institute (ISI) for his contributions to computational statistics. Dr. Liang has served as co-editor for the *Journal of Computational and Graphical Statistics* and as an associate editor for several other statistical journals. Dr. Liang has published two books and over 130 journal/conference papers covering a broad range of topics in data science, including uncertainty quantification for deep learning, dynamic learning, high-dimensional graphical modeling, high-dimensional statistical inference, big data computing, and Markov chain Monte Carlo. His research has also included interdisciplinary collaborations in the areas of biomedical science and engineering.

Dr. Bochao Jia is a Senior Advisor for Statistics-Immunology at Eli Lilly and Company, Indianapolis, Indiana, USA. He received his PhD degree in Biostatistics from the University of Florida in 2018. Dr. Jia is an expert on high-dimensional sparse graphical modeling and has authored quite a few papers on the topic.

List of Figures

List of Tables

Symbols

Symbol Description

$X_1, \ldots, X_p$ A set of random variables for which graphical models are to construct.

$\mathcal{P}_V$ The joint distribution of the random variables indexed by V.

$\mathbb{X}_1, \ldots, \mathbb{X}_n$ A set of *i.i.d* samples drawn from $\mathcal{P}_V$.

$G = (V, E)$ True conditional independence network, $V = \{1, 2, \ldots, p\}$ is the set of vertices or nodes, and $E = (e_{ij})$ is the adjacency matrix of the network.

$\hat{G}^{(n)} = (V, \hat{E}^{(n)})$ Estimate of G, where the superscript indicates the number of samples used in the estimation.

$\mathcal{G} = (\mathcal{V}, \mathcal{E})$ True correlation network, $\mathcal{V} = \{1, 2, \ldots, p\}$ is the set of vertices or nodes, and $\mathcal{E}$ is the adjacency matrix of the network.

$\mathcal{P}_V$ The joint distribution of the random variables indexed by V.

r_{ij} Correlation coefficient between the variables X_i and X_j.

$r_{z_{ij}}$ Correlation score between the variables X_i and X_j.

$\rho_{ij|V \setminus \{i,j\}}$ Partial correlation coefficient between the variables X_i and X_j conditioned on all other variables.

ψ_{ij} ψ-partial correlation coefficient between the variables X_i and X_j conditioned a neighboring set.

$\psi_{z_{ij}}$ ψ-score between the variables X_i and X_j.

$X \perp\!\!\!\perp Y | Z$ Variable X is independent of variable Y conditioned on the variable Z.

$|D|$ Cardinality of the set D.

$\|A\|_F$ The Frobenius norm of the matrix A.

$vec(A)$ The vectorization operator which converts the matrix A into a column vector.

$\xrightarrow{d}$ Convergence in distribution.

$\xrightarrow{p}$ Convergence in probability.

$\xrightarrow{a.s.}$ Almost sure convergence.

Chapter 1

Introduction to Sparse Graphical Models

This chapter provides a short introduction to high-dimensional graphical models, and introduces a unified paradigm of conditional independence tests for high-dimensional statistical inference as well.

1.1 Sparse Graphical Modeling

A graphical model [85], which usually refers to an undirected Markov network, has proven to be useful for the description of the dependency relationships among a large set of random variables. It is particularly useful under the high-dimensional scenario, i.e., when the number of random variables is much greater than the number of observations. In this scenario, with the aid of the sparse graphical model learned for the explanatory variables, the statistical inference for a high-dimensional regression model can be reduced to the statistical inference for a series of low-dimensional regression models, where the t-test or the Wald test can be used for inference of the significance and associated confidence interval for each explanatory variable. See Liang et al. [94], Sun and Liang [160] and Chapter 9 of this book for the development of this methodology.

A Bayesian network [128] can serve the same purpose as the graphical model for high-dimensional data analysis, but it is directed and focuses more on the data-generating mechanism. It is important to note that the Bayesian network can be constructed from its moral graph, an undirected Markov network, with intervention information of the data. For example, the collider set algorithm [130] or local neighborhood algorithm [111] can be applied for the purpose. On the other hand, given a Bayesian network, an undirected Markov network can be constructed by moralization. Since the intervention information is generally not available for the data collected in observation studies, the undirected Markov network is the focus of this book.

In practice, the observed data can be Gaussian, non-Gaussian or mixed, and the dependency between the associated random variables can be linear or nonlinear. Most works on graphical modeling focus on Gaussian graphical models, where the random variables are assumed to be Gaussian and the dependency between different random variables is assumed to be linear. Nonlinear and/or non-Gaussian graphical models

DOI: 10.1201/9780429061189-1

have also been studied in the recent literature. A brief review of the existing works on graphical modeling is provided as follows.

1.1.1 Gaussian Data

The methods of Gaussian graphical modeling are usually developed based on the relationships between the partial correlation coefficients, concentration matrix, and regression coefficients under the linear-Gaussian setting. To be more precise, let's consider a set of Gaussian random variables $\{X_1, X_2, \ldots, X_p\}$ drawn from $\mathcal{N}(\mathbf{0}, \Sigma)$, where Σ denotes the covariance matrix. Let $V = \{1, 2, \ldots, p\}$ denote the set of indices of the random variables, and let $\Theta = \Sigma^{-1}$ denote the concentration matrix. By the standard property of the multivariate Gaussian distribution, the conditional distribution of each variable X_j given all other variables is still a Gaussian distribution. More specifically, by equation (A.2), we have

$$X_j = \beta_{j,i} X_i + \sum_{r \in V \setminus \{i,j\}} \beta_{j,r} X_r + \varepsilon_j, \quad j = 1, 2, \ldots, p, \tag{1.1}$$

where ε_j is a zero-mean Gaussian random error and $\beta_{j,i}$ denotes a regression coefficient of the induced Gaussian linear regression. Further, by simple matrix calculation[1], we can show that

$$\rho_{ij|V \setminus \{i,j\}} = -\frac{\theta_{ij}}{\sqrt{\theta_{ii} \theta_{jj}}}, \tag{1.2}$$

where $\rho_{ij|V \setminus \{i,j\}}$ denotes the partial correlation coefficient between X_i and X_j given all other variables, and θ_{ij} is the $(i,j)^{th}$ entry of the matrix Θ. Also, the partial correlation coefficient $\rho_{ij|V \setminus \{i,j\}}$ is associated with the regression coefficient $\beta_{j,i}$ by the relation

$$\rho_{ij|V \setminus \{i,j\}} = \beta_{j,i} \frac{\sigma_{i|V \setminus \{i,j\}}}{\sigma_{j|V \setminus \{i,j\}}}, \tag{1.3}$$

where $\sigma_{i|V \setminus \{i,j\}}$ denotes the standard deviation of the residuals from the linear regression $X_i \sim X_{V \setminus \{i,j\}}$, and $\sigma_{j|V \setminus \{i,j\}}$ can be interpreted in the same way. Through this book, we will use X_A to denote a set of variables indexed by the set A, i.e., $X_A = \{X_i : i \in A\}$.

By summarizing the equations (1.1)–(1.3), we have

$$\rho_{ij|V \setminus \{i,j\}} \neq 0 \Longleftrightarrow \theta_{ij} \neq 0 \Longleftrightarrow \beta_{j,i} \neq 0. \tag{1.4}$$

Based on this relation, the graphical Lasso [6, 49, 193] infers the structure of the Gaussian graphical model by estimating the concentration matrix Θ through a regularization approach. Similarly, the Bayesian regularization method [54] estimates Θ under the Bayesian framework by letting each off-diagonal element of Θ be subject to a spike-and-slab Lasso prior [139]. The nodewise regression method [118] estimates

---

[1]See the Wikipedia article at `https://en.wikipedia.org/wiki/Partial_correlation#Using_matrix_inversion`, where the derivation is based on Schur's formula for block-matrix inversion.

the regression coefficients $\beta_{j,i}$'s through solving the p regressions in equation (1.1), where each is regularized by a Lasso penalty [164]. Similarly, Zheng et al. [201], estimated $\beta_{j,i}$'s through solving equation (1.1) from the perspective of structural equation modeling. The ψ-learning method [93] infers the structure of the Gaussian graphical model based on an equivalent measure of the partial correlation coefficient $\rho_{ij|V\setminus\{i,j\}}$, where the equivalent measure ψ_{ij} is defined such that

$$\psi_{ij} \neq 0 \Longleftrightarrow \rho_{ij|V\setminus\{i,j\}} \neq 0, \tag{1.5}$$

while ψ_{ij} can be calculated in a low-dimensional space, say, the space reduced by a sure independence screening procedure [45].

1.1.2 Non-Gaussian Data

Graphical modeling for non-Gaussian data is important, as many types of data generated in modern data science, such as single nucleotide polymorphisms (SNP) data, RNA sequencing (RNA-seq) data, and household incomes, are either discrete or non-Gaussian. Extending the graphical modeling methods for Gaussian data to non-Gaussian data has been an active research topic during the past decade. Some authors made the extension under the assumption that the random variables are still linearly dependent, although they can be discrete or non-normally distributed. For example, Ravikumar et al. [135], extended the nodewise regression method to binary data Cheng et al. [23], Lee and Hastie [88], and Yang et al. [186], extended the nodewise regression method to mixed data, Xu et al [183], extended the ψ-learning method to mixed data, and Fan et al. [44] extended the graphical Lasso method to mixed data by introducing some latent Gaussian random variables for the discrete data. Extension to nonlinear dependence has also been considered in the literature, see e.g., [101, 191, 202].

It is interesting to point out that some researchers found that nonlinearity and non-Gaussianity can actually be a blessing, which can reveal more accurate information about the data-generating mechanism than the linear and Gaussian approximation. For example Choi et al. [27], Hoyer et al. [68], and Shimizu et al. [147], found that non-Gaussianity is helpful in predicting causal relationships among the variables.

1.2 A Unified Paradigm for High-Dimension Statistical Inference

A fundamental issue underlying sparse graphical modeling is conditional independence test, which essentially requires one to simultaneously test the hypothesis

$$X_i \perp\!\!\!\perp X_j | X_{V\setminus\{i,j\}} \tag{1.6}$$

for each pair of variables (X_i, X_j) conditioned on all others. Under the high-dimensional scenario, an exact test for (equation 1.6) is generally unavailable. However, as shown in Liang and Liang [101], the test (equation 1.6) can be reduced to a low-dimensional test based on the sparsity of graphical models and the following simple mathematical fact on the conditional probability distribution:

$$
\begin{aligned}
P(X_i, X_j | X_{V \setminus \{i,j\}}) &= P(X_i | X_j, X_{V \setminus \{i,j\}}) P(X_j | X_{V \setminus \{i,j\}}) \\
&= P(X_i | X_j, X_{S_i^* \setminus \{j\}}) P(X_j | X_{S_{j \setminus i}^*}) \\
&= P(X_i, X_j | X_{S_i^* \cup S_{j \setminus i}^* \setminus \{j\}}) \\
&= P(X_i, X_j | X_{S_i \cup S_j \setminus \{i,j\}}),
\end{aligned}
\tag{1.7}
$$

where $X_{S_i^*}$ denotes the set of true variables of the nonlinear regression $X_i \sim X_{V \setminus \{i\}}$, $X_{S_{j \setminus i}^*}$ denotes the set of true variables of the nonlinear regression $X_j \sim X_{V \setminus \{i,j\}}$, S_i is a superset of S_i^*, and S_j is a superset of $S_{j \setminus i}^*$. Therefore, for each pair of variables (X_i, X_j), one only needs to find appropriate subsets S_i and S_j such that the test (equation 1.6) can be performed, the corresponding p-value or an equivalent measure of the p-value can be evaluated, and then the structure of the sparse graphical model can be determined through a multiple hypothesis test. This procedure will be detailed in the following chapters for different types of data under different assumptions. We note that the methods for the construction of the subsets S_i and S_j can be different under different assumptions for the distributions of X_i's and their dependency.

Other than graphical modeling, the mathematical fact (equation 1.7) can also be used in statistical inference for high-dimensional regression. As shown in Chapter 9, it reduces the high-dimensional inference problem to a series of low-dimensional inference problems, leading to more accurate confidence intervals and variable selection results compared to the regularization methods such as those developed by Javanmard and Montanari [71], van de Geer et al. [173], and Zhang and Zhang [196].

In summary, this book provides a unified paradigm of conditional independence tests for high-dimensional statistical inference, ranging from graphical modeling to variable selection and to uncertainty quantification.

1.3 Problems

1. Prove equation (1.2).

2. Prove equation (1.3).

Chapter 2

Gaussian Graphical Models

This chapter introduces the ψ-learning method [93]. Unlike the regularization methods such as graphical Lasso and nodewise regression, the ψ-learning method seeks to learn the Gaussian graphical model based on an equivalent measure of the partial correlation coefficient, which can be calculated in a lower dimensional space reduced through a sure independence screening procedure. With the equivalent measure of the partial correlation coefficient, the ψ-learning method has successfully converted the problem of high-dimensional Gaussian graphical modeling to a problem of multiple hypothesis conditional independence tests.

2.1 Introduction

Consider a dataset drawn from a multivariate Gaussian distribution $\mathcal{N}_p(\mu, \Sigma)$. Let $\mathcal{X} \in \mathbb{R}^{n \times p}$ denote the dataset, where n is the sample size, p is the dimension, and $n \ll p$ is generally assumed. Let X_j, $j \in V = \{1, 2, \ldots, p\}$, denote the random variables contained in the dataset, let $\boldsymbol{X}_j \in \mathbb{R}^n$ denote the jth column of $\mathcal{X}$ (i.e., n realizations of the variable X_j), and let $\boldsymbol{X}^{(i)} \in \mathbb{R}^p$ denote the ith row of $\mathcal{X}$ (i.e., i^{th} observation in the dataset). The general goal of Gaussian graphical modeling is to figure out the dependency relationships among the p random variables $X_1, X_2, \ldots, X_p$ under the assumption that these random variables interact pairwise and linearly with each other. Under this assumption, a celebrating property of the Gaussian graphical model is that it generally satisfies the relationships (equation 1.4). As briefly mentioned in Chapter 1, there have been quite a few methods developed in the literature by exploring these relationships from different perspectives.

In Gaussian graphical models, the dependency relationships among the p variables are represented by a graph or network denoted by $G = (V, E)$, where V denotes the set of nodes with each node representing a different Gaussian random variable, and $E = (e_{ij})$ denotes the adjacency matrix of the network. The value of $e_{ij} \in \{0, 1\}$ indicates the edge status between node i and node j. In particular, the absence of an edge indicates conditional independence given all other variables in the graph. For example, $e_{ij} = 0$ means $X_i \perp\!\!\!\perp X_j | \boldsymbol{X}_{V \setminus \{i, j\}}$, where $\boldsymbol{X}_A = \{X_k : k \in A\}$ denotes a set of variables indexed by the set A. This definition of conditional independence can be generalized to sets of random variables as follows: For any sets $I, J, U \subset V$, $X_I \perp\!\!\!\perp X_J | X_U$ denotes that X_I is conditionally independent of X_J given X_U. Refer to Appendix A.2 for more terms that are frequently used in the graphical model study.

DOI: 10.1201/9780429061189-2

2.2   Related Works

This section provides a brief review of some Gaussian graphical modeling methods by linking them with the relationships presented in equations (1.1)–(1.4).

2.2.1   Nodewise Regression

To estimate the Gaussian graphical model, Meinshausen and Bühlmann [118] took a straightforward way, estimating the regression coefficients $\beta_{j,i}$'s as defined in equation (1.1) through solving p high-dimensional regressions regularized by the Lasso penalty. More precisely, the nodewise regression method is to solve p-optimization problems

$$\hat{\beta}_{\backslash j} = \arg\min_{\beta_j}\{\|X_j - X_{V\backslash\{j\}}\beta_{\backslash j}\|_2 + \lambda_j\|\beta_{\backslash j}\|_1\}, \quad j = 1, 2, \ldots, p, \tag{2.1}$$

with each corresponding to a node in graph G, where λ_j is the regularization parameter, $\beta_{\backslash j} = (\beta_{j,1}, \beta_{j,2}, \ldots, \beta_{j,j-1}, \beta_{j,j+1}, \ldots, \beta_{j,p})^T$, and $\hat{\beta}_{\backslash j} = (\hat{\beta}_{j,1}, \hat{\beta}_{j,2}, \ldots, \hat{\beta}_{j,j-1}, \hat{\beta}_{j,j+1}, \ldots, \hat{\beta}_{j,p})^T$ is the regularized estimator of $\beta_{\backslash j}$. Let $\hat{C}_j = \{k : \hat{\beta}_{j,k} \neq 0\}$. Then the structure of graph G can be constructed with the "and"-rule:

$$\text{set } e_{jk} = e_{kj} = 1 \Longleftrightarrow k \in \hat{C}_j \text{ and } j \in \hat{C}_k,$$

or the "or"-rule:

$$\text{set } e_{jk} = e_{kj} = 1 \Longleftrightarrow k \in \hat{C}_j \text{ or } j \in \hat{C}_k.$$

The consistency of the method follows directly from the consistency of the Lasso regression under appropriate conditions such as the irrepresentable condition (for the data $\mathcal{X}$) and the beta-min condition (for $\min_{j,k}\{|\beta_{j,k}| : j,k \in V, j \neq k\}$) [199]. Obviously, the method also works with other types of penalties such as smoothly clipped absolute deviation (SCAD) [43], elastic net [206], and minimax concave penalty (MCP) [195] under appropriate conditions.

2.2.2   Graphical Lasso

Motivated by equation (1.4), the graphical Lasso method seeks to find a sparse estimate of Θ by minimizing the following regularized negative profile log-likelihood function:

$$-\log(\det(\Theta)) + tr(S\Theta) + \lambda\|vec(\Theta)\|_1, \tag{2.2}$$

where the mean vector μ has been replaced by its maximum likelihood estimator (MLE) $\hat{\mu} := \frac{1}{n}\sum_{i=1}^n X^{(i)}$, $S = \frac{1}{n}\sum_{i=1}^n (X^{(i)} - \hat{\mu})(X^{(i)} - \hat{\mu})^T$ is the empirical covariance matrix, and $vec(A)$ denote the vectorization operator which converts the matrix A to a column vector.

   Yuan and Lin [193] solved the minimization problem (2.2) using the interior point method [6], and Banerjee et al. [49], and Friedman et al. [175] solved the problem using the block coordinate descent algorithm [109].

Banerjee et al. [6] showed that problem (2.2) is convex and solved it by first estimating the covariance matrix $\Sigma = \Theta^{-1}$, rather than estimating Θ directly. Let W be an estimate of Σ. Partition the matrix W and S in the way:

$$W = \begin{pmatrix} W_{11} & w_{12} \\ w_{12}^T & w_{22} \end{pmatrix}, \quad S = \begin{pmatrix} S_{11} & s_{12} \\ s_{12}^T & s_{22} \end{pmatrix}.$$

Then Banerjee et al. [6] showed, by convex duality, that w_{12} can be estimated by solving the problem

$$\min_{\xi} \|W_{11}^{1/2}\xi - b\|_2^2 + \lambda\|\xi\|_1, \tag{2.3}$$

where $b = W_{11}^{-1/2}s_{12}/2$, and w_{12} and ξ are linked through $w_{12} = 2W_{11}\xi$. Banerjee et al. [6] solved problem (2.3) using an interior point optimization method, while [49] solved it using the fast coordinate descent algorithm [50]. The latter leads to a coordinate-wise update:

$$\xi^{(j)} \leftarrow \rho\left(s_{12}^{(j)} - 2\sum_{k \neq j} W_{11}^{(k,j)}\xi^{(k)}\right)/(2W_{11}^{(j,j)}), \quad j = 1, 2, \ldots, p-1, \tag{2.4}$$

where $\xi^{(j)}$ denotes the jth element of ξ, $W_{11}^{(k,j)}$ denotes the $(k, j)^{th}$ element of W_{11}, $s_{12}^{(j)}$ denotes the jth element of s_{12}, and $\rho(\cdot)$ is a soft-threshold operator

$$\rho(x,t) = \text{sign}(x)(|x| - t)_+.$$

An estimate W (for its non-diagonal elements only) can be obtained by cycling (2.4) through the columns of W, and a stable estimate W can be obtained by repeating this procedure until convergence. Finally, an estimate of Θ can be recovered from the estimate W cheaply.

2.2.3 Bayesian Regularization

Similar to graphical Lasso, the Bayesian regularization method [54] aims to find a sparse estimate of $\Theta = (\theta_{ij})$ but under the Bayesian framework. Gan et al. [54] imposed a mixture Laplace prior on the upper triangular entries θ_{ij}'s ($i < j$) and a weekly informative exponential prior on the diagonal entries θ_{ii}'s. Further, they constrained Θ to be positive definite, represented by $\Theta \succ 0$; and constrained the spectral norm of Θ to be upper bounded, represented by $\|\Theta\|_2 \leq B$ for some positive constant $B > 0$. Putting all these assumptions and constraints together leads to the following negative log-posterior distribution of Θ:

$$-\log \pi(\Theta|X^{(1)}, \ldots, X^{(n)}) = \text{Const.} + \frac{n}{2}\left(-\log(\det(\Theta)) + tr(S\Theta)\right)$$
$$+ \sum_{i<j} g_1(\theta_{ij}) + \sum_i g_2(\theta_{ij}), \tag{2.5}$$

where Θ takes values in the constrained matrix space $\{\tilde{\Theta} : \tilde{\Theta} \succ 0, \|\tilde{\Theta}\|_2 \leq B\}$,

$$g_1(\theta_{ij}) = -\log\left(\frac{\eta}{2v_1}e^{-\frac{|\theta_{ij}|}{v_1}} + \frac{1-\eta}{2v_0}e^{-\frac{|\theta_{ij}|}{v_0}}\right),$$

$$g_2(\theta_{ii}) = \tau|\theta_{ii}|,$$

and $0 < \eta < 1$, $v_1 > v_0 > 0$, and $\tau > 0$ are prior hyper-parameters.

Gan et al. [54] showed that if $B < (2nv_0)^{1/2}$, then the negative log-posterior $-\log \pi(\Theta|X^{(1)}, \ldots, X^{(n)})$ is convex over the constrained parameter space, which makes the method attractive in computation. Further, they proposed to minimize (equation 2.5) using the Expectation-Maximization (EM) algorithm [35] by introducing a binary latent variables R_{ij} for each of the upper triangular entries, where R_{ij} indicates the component of the mixture Laplace distribution that θ_{ij} belongs to *a priori*.

2.2.4   *Structural Equation Modeling*

Based on equation (1.1) Zheng et al. [201] proposed a structural equation modeling method for learning a directed acyclic graph (DAG) for $X_1, X_2, \ldots, X_p$. The method aims to find a sparse coefficient matrix $\beta = (\beta_{j,i}) \in \mathbb{R}^{p \times p}$ that solves equation (1.1) by minimizing a regularized objective function

$$\frac{1}{2p}\|X - X\beta\|_F^2 + \frac{\rho}{2}|h(\beta)|^2 + \alpha h(\beta) + \lambda\|vec(\beta)\|_1, \qquad (2.6)$$

where ρ, α and λ are tuning parameters, $h(\beta) = tr(e^{\beta \circ \beta}) - p^2$, $\circ$ is the Hadamard product, and e^A is the matrix exponential of A. The penalty $\|vec(\beta)\|_1$ induces the sparsity for β, and the two penalty terms of $h(\beta)$ impose an acyclicity constraint on the graph and ensure a DAG to be learned. The key insight of the method is to replace the combinatorial constraint that is necessary for learning a DAG by the differentiable constraint $h(\beta) = 0$ such that a gradient-based optimization algorithm can be applied. Given the DAG, an undirected Markov network can then be constructed by moralization as described in Chapter 6.

2.3 ψ-Learning Method

Motivated by equation (1.4), the ψ-learning method [93] aims to find an equivalent measure of $\rho_{ij|V\setminus\{i,j\}}$, denoted by ψ_{ij}, such that it can be computed under the scenario $n \ll p$, while being equivalent with $\rho_{ij|V\setminus\{i,j\}}$ in Gaussian graphical modeling in the sense that $\rho_{ij|V\setminus\{i,j\}} = 0$ if and only if $\psi_{ij} = 0$.

To construct such an equivalent measure, we need to make some assumptions about Gaussian graphical models. Let $\mathcal{P}_V$ denote the joint distribution of the variables $\{X_k : k \in V\}$, and let P_A denote the marginal distribution associated with the random variables in $A \subseteq V$. In particular, we assume that the Gaussian graphical models satisfy the Markov property and faithfulness conditions, which are defined as follows.

Definition 2.1 (Markov property) *$\mathcal{P}_V$ is said to satisfy the Markov property with respect to graph G if for every triple of disjoint sets $I, J, U \subset V$, it holds that $X_I \perp\!\!\!\perp X_J | X_U$ whenever U is a separator of I and J in G.*

Definition 2.2 (Faithfulness) *$\mathcal{P}_V$ is said to satisfy the faithfulness condition with respect to graph G: For every triple of disjoint sets $I, J, U \subseteq V$, U separates I and J in G if and only if $X_I \perp\!\!\!\perp X_J | X_U$ holds.*

The *separator* is defined in Appendix A.2. The concept of faithfulness has been well discussed in Meek [113] and Sportes et al. [152]. For the Gaussian graphical model, with respect to the Lebesgue measure over its parameter space, the set of multivariate Gaussian distributions that are unfaithful to graph G is measure zero. See Theorem 8 of Meek [113] for a formal theoretical justification for this statement.

2.3.1 ψ-Partial Correlation Coefficient

Let r_{ij} denote the correlation coefficient of X_i and X_j, and let $\mathcal{G} = (\mathcal{V}, \mathcal{E})$ denote the correlation network of $X_1, \ldots, X_p$, where $\mathcal{V} = V$, and $\mathcal{E} = (\tilde{e}_{ij})$ is the associated adjacency matrix. Let $\hat{r}_{ij}$ denote the empirical correlation coefficient of X_i and X_j, let γ_i denote a threshold value, and let $\hat{\mathcal{E}}_{\gamma_i,i}^{(n)} = \{v : |\hat{r}_{iv}| > \gamma_i\}$ denote the neighboring set of node i in the empirical correlation network, where the superscript "(n)" indicates the sample size of the dataset. Similarly, we define $\hat{\mathcal{E}}_{\gamma_j,j}^{(n)} = \{v : |\hat{r}_{jv}| > \gamma_j\}$, $\hat{\mathcal{E}}_{\gamma_i,i,-j}^{(n)} = \{v : |\hat{r}_{iv}| > \gamma_i\} \setminus \{j\}$, and $\hat{\mathcal{E}}_{\gamma_j,j,-i}^{(n)} = \{v : |\hat{r}_{jv}| > \gamma_j\} \setminus \{i\}$.

For a pair of nodes X_i and X_j, Liang et al. [93] defined the ψ-partial correlation coefficient as

$$\psi_{ij} = \rho_{ij|S_{ij}}, \tag{2.7}$$

where S_{ij} is the smaller of the two sets $\hat{\mathcal{E}}_{\gamma_i,i,-j}^{(n)}$ and $\hat{\mathcal{E}}_{\gamma_j,j,-i}^{(n)}$. Since the size of the conditioning set S_{ij} is generally much smaller than the sample size n, ψ_{ij} can be calculated according to equation (1.2) through inverting the covariance matrix of the variables $\{X_k : k \in S_{ij} \cup \{i, j\}\}$. Figure 2.1 illustrates the calculation of ψ_{ij} based on a correlation graph.

Let $n_G(i)$ denote the set of neighboring nodes of X_i in graph G, i.e., $n_G(i) = \{k : e_{ik} = 1, k \in V\}$. Let G_{-ij} be a reduced graph of G, where the edge between the nodes X_i and X_j, if existing, is deleted. Theorem 2.1 shows that ψ_{ij} and $\rho_{ij|V\setminus\{i,j\}}$ are equivalent in the sense of equation (2.8) and thus they are equivalent for graph recovery.

Theorem 2.1 *(Theorem 1; [93]) Suppose that a Gaussian graphical model $G = (V, E)$ satisfies the Markov property and faithfulness conditions, and $n_G(i) \subseteq \hat{\mathcal{E}}_{\gamma_i,i}^{(n)}$ holds for each node X_i. Then ψ_{ij} defined in equation (2.7) is an equivalent measure of the partial correlation coefficient $\rho_{ij|V\setminus\{i,j\}}$ in the sense that*

$$\psi_{ij} = 0 \iff \rho_{ij|V\setminus\{i,j\}} = 0. \tag{2.8}$$

Proof 1 *If $\psi_{ij} = 0$, then $\rho_{ij|V\setminus\{i,j\}} = 0$ by the Markov property. On the other hand, if $\rho_{ij|V\setminus\{i,j\}} = 0$ holds, then $e_{ij} = 0$ by faithfulness. In this case, it follows from the*

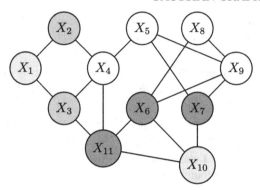

Figure 2.1 *A correlation graph $\mathcal{G} = (\mathcal{V}, \mathcal{E})$ of Gaussian random variables $X_1, X_2, \ldots, X_{11}$, where the green nodes $\{X_2, X_3\}$ form the neighborhood of X_1, and the red nodes $\{X_6, X_7, X_{11}\}$ form the neighborhood of X_{10}. Given the correlation graph $\mathcal{G}$, we have $\psi_{1,10} = \rho_{1,10|2,3}$, i.e., the partial correlation coefficient of X_1 and X_{10} given $\{X_2, X_3\}$.*

assumption $n_G(i) \subseteq \hat{\mathcal{E}}_{\gamma_i, i}$ *that* S_{ij} *forms a separator for the nodes* X_i *and* X_j *in the reduced graph* G_{-ij} *and thus, by the Markov property,* $\psi_{ij} = 0$ *holds.*

As implied by the proof, the conditioning set S_{ij} that can be used in calculation of ψ_{ij} is not unique. For example, one can set S_{ij} to any of the sets, $\hat{\mathcal{E}}_{\gamma_i, i, -j}^{(n)}$, $\hat{\mathcal{E}}_{\gamma_j, j, -i}^{(n)}$, or $\hat{\mathcal{E}}_{\gamma_i, i, -j}^{(n)} \cup \hat{\mathcal{E}}_{\gamma_j, j, -i}^{(n)}$, while not alternating the equivalence property (equation 2.8). However, as implied by Fisher's z-transformation, see equation (2.9), the variance of ψ_{ij} can increase with the size of S_{ij} and, therefore, it is generally preferred to set S_{ij} to the smaller of $\hat{\mathcal{E}}_{\gamma_i, i, -j}^{(n)}$ and $\hat{\mathcal{E}}_{\gamma_j, j, -i}^{(n)}$. Note that the setting $S_{ij} = \hat{\mathcal{E}}_{\gamma_i, i, -j}^{(n)} \cup \hat{\mathcal{E}}_{\gamma_j, j, -i}^{(n)}$ matches the formula (1.7), which is derived without the Markovian and faithfulness assumptions. This further implies that the Markovian and faithfulness assumptions potentially enable the Gaussian graphical models more accurately recovered.

2.3.2 ψ-Learning Algorithm

By equation (1.4), one has $\beta_{i,k} \neq 0 \iff k \in n_G(i)$ for any $k \neq i$. Further, by the marginal correlation screening property of linear regression [45], the condition $n_G(i) \subseteq \hat{\mathcal{E}}_{\gamma_i, i}^{(n)}$ of Theorem 2.1 almost surely holds as the sample size $n \to \infty$. Motivated by this observation, Liang et al. [93] proposed the following algorithm for estimating high-dimensional Gaussian graphical models.

Algorithm 2.1 *(ψ-Learning; [93])*

a. *(Correlation screening) Determine the neighboring set $\hat{\mathcal{E}}_{\gamma_i, i}^{(n)}$ for each node X_i.*

    i. *Conduct a multiple hypothesis test for the correlation coefficients $\{r_{ij}\}$ at a significance level of α_1 to identify the structure of the correlation network.*

    ii. *For each node X_i, identify its neighboring set in the empirical correlation network, and reduce the size of the neighboring set to $n/[\xi_n \log(n)]$ by removing those with smaller values of $|\hat{r}_{ij}|$'s.*

b. *(ψ-calculation) For each pair of nodes X_i and X_j, find the conditioning set S_{ij} based on the empirical correlation network and calculate the ψ-partial correlation coefficient ψ_{ij}.*

c. *(ψ-screening) Conduct a multiple hypothesis test for the ψ-partial correlation coefficients $\{\psi_{ij} : i, j \in V, i \neq j\}$ at a significance level of α_2 to identify the structure of the Gaussian graphical model.*

To conduct a multiple hypothesis test required in steps (a) and (c), Liang et al. [93] suggested the empirical Bayesian method [95]. In this method, they first applied Fisher's z-transformation to transform the correlation coefficients and ψ-partial correlation coefficients to z-scores, calculated the p-values for the individual hypothesis tests, and then performed the multiple hypothesis tests as described in Appendix A.7. To be more detailed, for step (a), they made the transformations:

$$z_{ij} = \frac{1}{2} \log \left[\frac{1 + \hat{r}_{ij}}{1 - \hat{r}_{ij}} \right], \quad i, j \in V, i \neq j.$$

Under the null hypothesis $H_0 : r_{ij} = 0$, z_{ij} is approximately distributed according to the Gaussian distribution $N(0, \frac{1}{n-3})$, and thus a p-value can be calculated for the individual test $H_0 : r_{ij} = 0 \leftrightarrow H_1 : r_{ij} \neq 0$. Similarly, for step (c), they made the transformations:

$$z'_{ij} = \frac{1}{2} \log \left[\frac{1 + \hat{\psi}_{ij}}{1 - \hat{\psi}_{ij}} \right], \quad i, j \in V, i \neq j, \tag{2.9}$$

where z'_{ij} is approximately distributed according to $N(0, \frac{1}{n-|S_{ij}|-3})$ under the null hypothesis $H_0 : \psi_{ij} = 0$.

Obviously, other multiple hypothesis testing methods, such as those developed in Benjamini et al. [10] and Efron [41] can also be used in Algorithm 2.1. We particularly pointed out that all these methods have effectively accounted for the dependence between test statistics.

Algorithm 2.1 contains three parameters ξ_n, α_1, and α_2 to be specified by the user. The parameter ξ_n should be set in the order of $O(1)$, while ensuring $n/[\xi_n \log(n)] < n - 4$ holds. The parameter α_1 can be set to a slightly large value, e.g., 0.05, 0.1, 0.2, or even 0.25, which increases the likelihood that $n_G(i) \subseteq \hat{\mathcal{E}}_{\gamma_1, i}$ holds for each node X_i as required by Theorem 2.1. The parameter α_2 can be set according to our targeted sparsity level, e.g., 0.01 or 0.05, of the graphical model. In general, a smaller value of α_2 will lead to a sparser graph.

2.4 Simulation Studies

The ψ-learning algorithm has been implemented in the R package *equSA*[1] [76], whose name is coined to reflect the two major components of the algorithm, equivalent measure of the partial correlation coefficient and the <u>S</u>tochastic <u>A</u>pproximation algorithm [137] used in the empirical Bayesian multiple hypothesis tests. With this package, we simulated data from Gaussian graphical models with

---

[1]The package is available at https://cran.r-project.org/src/contrib/Archive/equSA/ and Dr. Liang's homepage https://www.stat.purdue.edu/~fmliang/.

different types of structures, including band, random, cluster, and scale-free. In particular, for the band structure, we specify the concentration matrix $\Theta = (\theta_{ij})$ as

$$
\theta_{ij} = \begin{cases} 1, & \text{if } i = j, \\ 0.5, & \text{if } |j - i| = 1, \\ 0.25, & \text{if } |j - i| = 2, \\ 0, & \text{otherwise,} \end{cases} \tag{2.10}
$$

which is also called the AR(2) structure in the package *equSA*. Refer to *equSA* for the detailed setup for the other structures. For each structure, we simulated 10 datasets independently. Each dataset consisted of $n = 200$ samples and $p = 300$ variables. Figure 2.2 shows the network structures of the simulated Gaussian graphical models.

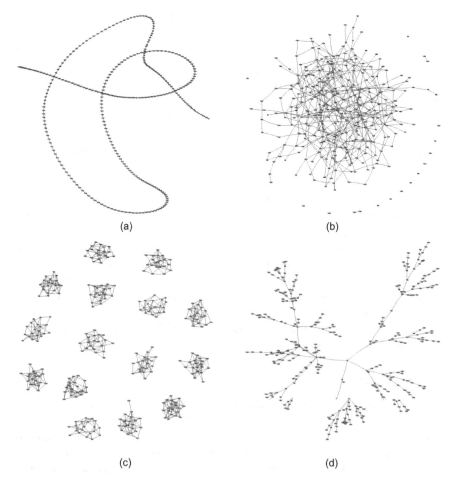

Figure 2.2 *Network structures of the simulated Gaussian graphical models, which are called band (a), random (b), cluster (c), and scale-free (d), respectively.*

Table 2.1 *Average AUC values (over ten datasets) produced by ψ-learning, graphical Lasso (gLasso), and nodewise regression for the simulated Gaussian graphical models with different structures, where "SD" represents the standard deviation of the average AUC value, and the p-value is for a paired t-test with the hypothesis $H_0 : AUC_\psi = AUC_\phi$ versus $H_1 : AUC_\psi > AUC_\phi$, and ϕ represents graphical Lasso or nodewise regression.*

Structure	Measure	gLasso	Nodewise	ψ-learning
	ave-AUC	0.6717	0.8225	0.9778
AR(2)	SD	(0.0024)	(0.0023)	(0.0019)
	p-value	<2.20e-16	8.66e-15	—
	ave-AUC	0.8364	0.8433	0.8597
Random	SD	(0.0108)	(0.0116)	(0.0100)
	p-value	3.81e-5	3.48e-4	—
	ave-AUC	0.5978	0.6378	0.6646
Cluster	SD	(0.0106)	(0.0107)	(0.0130)
	p-value	3.53e-8	5.05e-5	—
	ave-AUC	0.8274	0.8349	0.8695
Scale free	SD	(0.0106)	(0.0117)	(0.0077)
	p-value	1.85e-6	1.81e-5	—

The ψ-learning algorithm was compared with graphical Lasso[2] [49] and node-wise regression[3] [118] on the simulated data. The numerical results are summarized in Table 2.1, which reports the average areas under the precision-recall curves (abbreviated as "AUC") produced by each method. Refer to Appendix A.4 for the definition of the precision-recall curve. For ψ-learning, we calculated the AUC values by varying the significance level α_2, while fixing $\alpha_1 = 0.05$. For graphical Lasso and nodewise regression, we calculated the AUC values by varying their regularization parameters. Refer to Appendix A.8 for the code used in producing some values of Table 2.1.

The comparison shows that the ψ-learning algorithm significantly outperforms the graphical Lasso and nodewise regression, although the significance depends on the network structure. In particular, for AR(2), the improvement is drastic. For this structure, the neighboring set $n_G(i)$ of each node X_i is small, and thus the neighboring set screening condition $n_G(i) \subseteq \hat{\mathcal{E}}_{\gamma,i}^{(n)}$ in Theorem 2.1 can be easily satisfied. For the other structures, the neighboring sets of some nodes might be large and thus the neighboring set screening condition might be violated, which can adversely affect the performance of the method.

To give more explanations for the performance of the ψ-learning method, we plot in Figure 2.3 the histogram of the ψ-scores for a dataset simulated with the AR(2) structure, where the ψ-score is given by

---

[2]It is implemented in the R package *huge* [200].
[3]It is implemented in the R package *huge* [200].

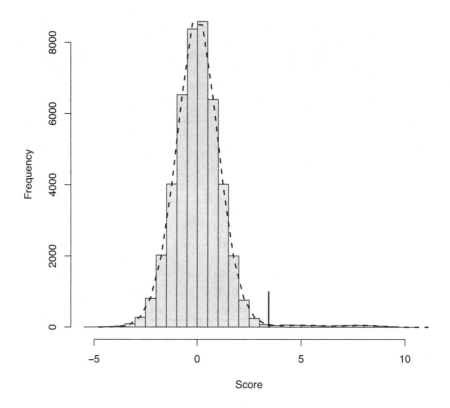

Figure 2.3 *Histogram of the ψ-scores for a dataset simulated with the AR(2) structure, where the dashed curve represented the density function of the fitted exponential power distribution for the non-significant ψ-scores, and the small bar represents the cutoff value for the significant ψ-scores.*

$$\hat{\Psi}_{ij} = \Phi^{-1}\left(2\Phi(\sqrt{n - |S_{ij}|} - 3|z'_{ij}|) - 1\right),$$

and z'_{ij} is as defined in equation (2.9). The cutoff value for the significant ψ-scores is determined using the empirical Bayesian method [95] by setting Storey's q-value [156] to $\alpha_2 = 0.05$. The ψ-learning method controls the false discovery rate (FDR) of the identified edges (i.e., significant ψ-scores) by choosing an appropriate value of α_2.

2.5   An Application for Gene Regulatory Network Recovery

The dataset[4] contains the expression levels of $p = 3883$ genes measured in $n = 49$ samples from different breast cancer patients [179]. The full graph for this example

---

[4]The dataset is available at http://strimmerlab.org/data.html.

consists of 7,536,903 edges. This example has been analyzed by Liang et al. [93], and we summarized their results as follows.

Figure 2.5a shows the gene regulatory network (GRN) recovered by the ψ-learning algorithm at a significance level of $\alpha_2 = 0.01$, and Figure 2.4 shows the edges of the network. With this network, a few hub genes, i.e., those with strong connectivity to other genes, can be identified. As reported by Liang et al. [93], the top four hub genes include CD44, IGFBP-5, HLA, and STARD3, which are all associated with the molecular mechanism of breast cancer. In particular, CD44 is a biomarker gene for breast cancer stem cells [107], IGFBP-5 is a potential therapeutic target for breast cancer [1], and HLA and STARD3 are associated with the recurrent risk and poor clinical outcome of breast cancer [79].

Figure 2.5b and c show, respectively, the gene regulatory networks identified by graphical Lasso and nodewise regression for the example. Unfortunately, both networks are empty, which might be due to the small sample size of this dataset. In our experience, the regularization methods often need a relatively large dataset for producing reliable results, see e.g., [98] for some discussions on this issue. Compared to graphical Lasso and nodewise regression, the ψ-learning algorithm can work with a much smaller sample size due to its dimension reduction nature.

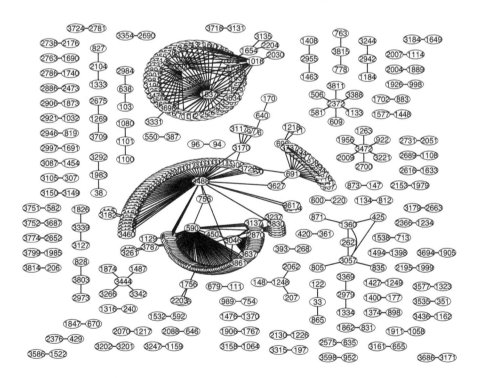

Figure 2.4 *Edges of the GRN learned by the ψ-learning algorithm with the breast cancer data, where the genes numbered by 590, 3488, 1537, and 992 are CD44, IGFBP-5, HLA, and STARD3, respectively [93].*

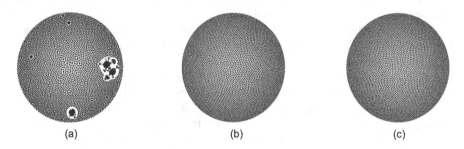

(a)   (b)   (c)

Figure 2.5 *Gene regulatory networks learned with the breast cancer data by (a) ψ-learning, (b) graphical Lasso, and (c) nodewise regression [93].*

2.6 ψ-Learning with Extra Data

In practice, we have often the case that other than the target dataset, some extra and related datasets are also available. The ψ-learning algorithm is very flexible for making use of the extra datasets for possible tasks such as data integration, network comparison, and covariate effect adjustment. Following Liang et al. [93], we make the following descriptions.

2.6.1 Data Integration

To make use of the wealth of data generated in modern data science to improve the accuracy of statistical inference, data integration has been a natural strategy. However, for many existing sparse graphical modeling methods, this is not obvious due to their point estimation nature, which usually fails to provide any uncertainty measure for their graphical model estimate.

For the ψ-learning algorithm, this is straightforward since it has broken the task of graphical modeling into a series of edgewise conditional independence tests. Then data integration can be done based on the p-values of the conditional independence tests from different sources of data with some existing meta-analysis methods such as Stouffer's Z-test method [157, 194] and Karl Pearson's p-value combination method (see e.g. [126]). In the case that the data of different sources are dependent, the meta-analysis methods, e.g., the harmonic mean p-value method [180], Brown's method [16, 132], and Kost's method [82], can be used, which are all developed for dependent p-values.

In what follows, we describe Stouffer's meta-analysis method for illustration of the ψ-integration method. In this method, we can first transform the ψ-partial correlation coefficients to Z-scores through Fisher's z-transformation:

$$Z_{\psi_{ij}} = \frac{\sqrt{n - |S_{ij}| - 3}}{2} \log \left[\frac{1 + \hat{\psi}_{ij}}{1 - \hat{\psi}_{ij}} \right], \quad i, j = 1, 2, \ldots, p, \qquad (2.11)$$

then we can combine the Z-scores from different sources of data by

$$\bar{Z}_{\psi_{ij}} = \frac{\sum_{k=1}^{K} w_k Z_{\psi_{ij}}^{(k)}}{\sqrt{\sum_{k=1}^{K} w_k^2}}, \quad i,j = 1,2,\ldots,p, \tag{2.12}$$

where K denotes the total number of data sources, $Z_{\psi_{ij}}^{(k)}$ denotes the Z-score from source k, and w_k denotes the weight assigned on source k. If we assume that $\{Z_{\psi_{ij}}^{(k)} : k = 1,2,\ldots,K\}$ are mutually independent, then $\bar{Z}_{\psi_{ij}}$ follows the standard Gaussian distribution and a multiple hypothesis test can be further done on $\bar{Z}_{\psi_{ij}}$'s for identifying the structure of the combined Gaussian graphical model.

To illustrate the ψ-integration procedure, we simulated two independent datasets from a graphical model with a hub structure. The model consists of $p = 200$ nodes, which are evenly divided into ten disjoint hub groups. In our simulations, we set the sample size $n = 100$ for the first dataset and $n = 75$ for the second dataset. Figure 2.6 shows the networks learned by ψ-learning for each dataset and that learned by the ψ-integration procedure. The comparison indicates that the ψ-integration procedure has led to a much improved estimate for the underlying true network.

2.6.2 Network Comparison

Network comparison is important, which, for example, can help us to understand the cancer molecular working mechanism by comparing the gene regulatory networks of the cancer and normal tissues. However, like for data integration, how to compare different networks is not obvious for many existing sparse graphical modeling methods.

In contrast, network comparison is straightforward for the ψ-learning algorithm based on its edgewise conditional independent tests. With those tests, ψ-learning

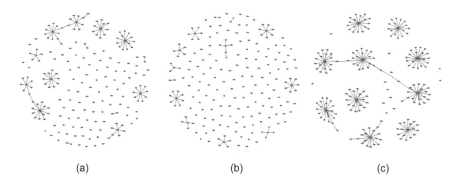

(a)   (b)   (c)

Figure 2.6 *Illustration of the ψ-integration method: (a) the network identified by the ψ-learning method for the dataset with $n = 100$ and $p = 200$; (b) the network identified by the ψ-learning method for the dataset with $n = 75$ and $p = 200$; (c) the network produced by the ψ-integration method.*

converts the network comparison problem to a multiple hypothesis test: simultaneously test the hypotheses

$$H_0^{(i,j)} : e_{ij}^{(1)} = e_{ij}^{(2)} \leftrightarrow H_1^{(i,j)} : e_{ij}^{(1)} \neq e_{ij}^{(2)}, \quad \text{for } i, j \in V \text{ and } i < j,$$

where $e_{ij}^{(k)}$ indicates the edge status of the network under condition $k \in \{1,2\}$. The test statistic is given by

$$Z_{d_{ij}} = [Z_{\psi_{ij}}^{(1)} - Z_{\psi_{ij}}^{(2)}]/\sqrt{2}, \tag{2.13}$$

where $Z_{\psi_{ij}}$ is a Z-score as defined in equation (2.11), and $Z_{d_{ij}}$ is distributed as the standard Gaussian distribution under the null hypothesis.

To illustrate the network comparison procedure, we simulated two datasets from the graphical models with a scale-free structure and a hub structure, respectively. For each dataset, we set $n = 200$ and $p = 200$. Figure 2.7 shows the networks learned by ψ-learning for the two datasets as well as the edges that are identified by the procedure to be significantly different in the two networks.

Extension of the pairwise network comparison to multiple network comparison is straightforward, for which a multiple comparison procedure, e.g., Fisher's least significance difference test, can be adopted for the comparison.

2.6.3   Covariate Effect Adjustment

Covariate effect adjustment is important for Gaussian graphical model construction, as it can be affected by some external confounding factors, see e.g. [19, 190] for more discussions on this issue.

For the ψ-learning method, the covariate effect can be simply adjusted by expanding the conditioning set of each ψ-partial correlation coefficient to include the external factors, if the external factors also follow Gaussian distributions. Otherwise, we can replace the ψ-partial correlation coefficient by a p-value from a conditional

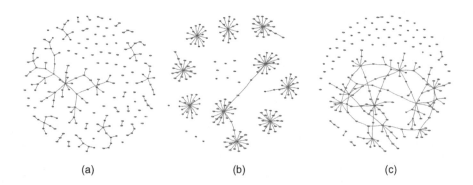

(a)   (b)   (c)

Figure 2.7 *Illustration of the network comparison method: (a) the network identified by the ψ-learning method for the model with the scale-free structure; (b) the network identified by the ψ-learning method for the model with the hub structure; and (c) the edges that are identified by the method to be significantly different in the two networks.*

independence test. More precisely, in Algorithm 2.1, the ψ-partial correlation coefficient can be replaced by the p-value from the test $H_0 : \beta_{q+1} = 0 \leftrightarrow H_1 : \beta_{q+1} \neq 0$ in the regression

$$X_i = \beta_0 + \beta_1 W_1 + \cdots + \beta_q W_q + \beta_{q+1} X_j + \sum_{k \in S_{ij}} \gamma_k X_k + \varepsilon, \qquad (2.14)$$

where $W_1, W_2, \ldots, W_q$ denote the external covariates, ε is Gaussian random error, and S_{ij} denotes the conditioning set obtained from the correlation screening step.

In correlation screening, the effect of the external factors can be accounted for through the following regression:

$$X_i = \beta_0 + \beta_1 W_1 + \cdots + \beta_q W_q + \beta_{q+1} X_j + \varepsilon, \qquad (2.15)$$

through testing the hypotheses $H_0 : \beta_{q+1} = 0 \leftrightarrow H_1 : \beta_{q+1} \neq 0$. However, this is not necessary, although this might lead to slightly larger conditioning sets of S_{ij}'s in equation (2.14).

2.7 Consistency of the ψ-Learning Algorithm

This section provides a theoretical guarantee for the consistency of the ψ-learning algorithm under the high-dimensional regime that p is much greater than n and, in addition, p can grow with n at an exponential rate. To indicate this dependency of p on n, we rewrite p, $\mathcal{P}_V$, $G = (V, E)$, and $\mathcal{G} = (\mathcal{V}, \mathcal{E})$ as p_n, $\mathcal{P}_V^{(n)}$, $G^{(n)} = (V^{(n)}, E^{(n)})$, and $\mathcal{G}^{(n)} = (\mathcal{V}^{(n)}, \mathcal{E}^{(n)})$, respectively. Note that $\mathcal{V}^{(n)} = V^{(n)}$, $E^{(n)} = \{(i, j) : \rho_{ij|V \setminus \{i,j\}} \neq 0, \, i, j = 1, \ldots, p_n\}$, and $\mathcal{E}^{(n)} = \{(i, j) : r_{ij} \neq 0, \, i, j = 1, \ldots, p_n\}$.

Let $\hat{r}_{ij}$ denote the empirical correlation coefficient of X_i and X_j, and let γ_n denote a threshold of empirical correlation coefficient used in the correlation screening step of Algorithm 2.1. Define

$$\hat{\mathcal{E}}_{\gamma_n}^{(n)} = \{(i, j) : |\hat{r}_{ij}| > \gamma_n\}, \quad \text{and} \quad \hat{\mathcal{E}}_{\gamma_n, i}^{(n)} = \{j : j \neq i, |\hat{r}_{ij}| > \gamma_n\}. \qquad (2.16)$$

Let ζ_n denote the threshold of ψ-partial correlation coefficient used in the ψ-screening step of Algorithm 2.1. Similar to equation (2.16), define

$$\hat{E}_{\zeta_n}^{(n)} = \{(i, j) : |\hat{\psi}_{ij}| > \zeta_n, \, i, j = 1, 2, \ldots, p_n\}, \qquad (2.17)$$

which denotes the edge set of the Gaussian graphical model identified through thresholding the ψ-partial correlation coefficients.

In what follows, we follow [93] to give the assumptions and main intermediate results for proving the consistency of the ψ-learning algorithm, while omitting the detailed proofs. We refer the readers to [93] for the detailed proofs. The whole strategy of the proof follows [78] in the analysis of the PC algorithm [153], and the proofs of Lemmas 2.1 and 2.2 are slightly modified from Luo et al. [108].

Assumption 2.1 $\mathcal{P}_V^{(n)}$ *is multivariate Gaussian, and it satisfies the Markov property and faithfulness condition with respect to the undirected graph $G^{(n)}$ for all $n \in \mathbb{N}$.*

Assumption 2.2 $p_n = O(\exp(n^\delta))$ *for some constant* $0 \leq \delta < 1$.

The faithfulness condition in Assumption 2.1 is generally required for the conditional independence tests-based methods, see e.g. the PC algorithm [78, 153]. We note that this condition can be slightly weakened to the adjacency faithfulness condition as in Liang et al. [93]. Assumption 2.2 allows the dimension p_n to increase with the sample size n in an exponential rate.

Assumption 2.3 *The correlation coefficients satisfy the inequalities:*

$$\min\{|r_{ij}|; r_{ij} \neq 0, \, i, j = 1, 2, \ldots, p_n, \, i \neq j\} \geq c_0 n^{-\kappa}, \qquad (2.18)$$

for some constants $c_0 > 0$ *and* $0 < \kappa < (1 - \delta)/2$, *and*

$$\max\{|r_{ij}|; i, j = 1, \ldots, p_n, i \neq j\} \leq M_r < 1, \qquad (2.19)$$

for some constant $0 < M_r < 1$.

Assumption 2.3 ensures detectability of nonzero correlation coefficients by imposing a lower bound on $\min\{|r_{ij}| : i, j = 1, 2, \ldots, p_n, i \neq j\}$, and avoids the non-identifiability (up to a linear transformation) issue by imposing an upper bound on $\max\{|r_{ij}| : i, j = 1, 2, \ldots, p_n, i \neq j\}$. Further, by the correlation screening property of linear regression [45], there exist constants $c_1 > 0$ and $0 < \kappa' \leq \kappa$ such that

$$\min\{|r_{ij}|; (i, j) \in E^{(n)}, i, j = 1, \ldots, p_n\} \geq c_1 n^{-\kappa'}. \qquad (2.20)$$

Assumption 2.4 *There exist constants* $c_4 > 0$ *and* $0 \leq \tau < 1 - 2\kappa'$ *such that* $\lambda_{\max}(\Sigma^{(n)}) \leq c_4 n^\tau$, *where* $\Sigma^{(n)}$ *denotes the covariance matrix of* $(X_1, X_2, \ldots, X_p)$, *and* $\lambda_{\max}(\Sigma^{(n)})$ *denotes the largest eigenvalue of* $\Sigma^{(n)}$.

Assumption 2.4 restricts the growth rate of the largest eigenvalue of $\Sigma^{(n)}$, which ensures sparsity of the thresholded correlation network as implied by Lemma 2.3.

Assumption 2.5 *The* ψ-*partial correlation coefficients satisfy the inequality:* $\inf\{\psi_{ij}; \psi_{ij} \neq 0, \, i, j = 1, \ldots, p_n, \, i \neq j, \, |S_{ij}| \leq q_n\} \geq c_5 n^{-d}$, *where* $q_n = O(n^{2\kappa' + \tau})$, $0 < c_5 < \infty$ *and* $0 < d < (1 - \delta)/2$ *are some constants. In addition,* $\sup\{\psi_{ij}; i, j = 1, \ldots, p_n, \, i \neq j, |S_{ij}| \leq q_n\} \leq M_\psi < 1$, *for some constant* $0 < M_\psi < 1$.

Similar to Assumptions 2.3 and 2.5 ensures detectability of nonzero ψ-partial correlations and avoids the non-identifiability issue of the variables by imposing an upper bound M_ψ on ψ_{ij}'s.

Lemma 2.1 *(Lemma 1; [93]) Suppose Assumptions 2.1–2.3 hold. Let* $\gamma_n = 2/3 c_1 n^{-\kappa'}$. *Then there exist constants* c_2 *and* c_3 *such that*

$$P(E^{(n)} \subseteq \hat{\mathcal{E}}^{(n)}_{\gamma_n}) \geq 1 - c_2 \exp(-c_3 n^{1-2\kappa'}),$$

$$P(b_{G^{(n)}}(i) \subseteq \hat{\mathcal{E}}^{(n)}_{\gamma_n, i}) \geq 1 - c_2 \exp(-c_3 n^{1-2\kappa'}).$$

Lemma 2.1 establishes the sure screening property of the thresholded correlation network: As $n \to \infty$, the probability that the true Gaussian graphical model $E^{(n)}$ is contained in the thresholded correlation network tends to 1. Therefore, the ψ-partial correlation coefficients can be calculated by conditioning on the neighborhood sets identified in the thresholded correlation network.

Lemma 2.2 *(Lemma 2; [93]) Suppose Assumptions 2.1–2.4 hold. Let $\gamma_n = 2/3c_1 n^{-\kappa'}$. Then for each node i,*

$$P\left[|\hat{\mathcal{E}}_{\gamma_n,i}^{(n)}| \leq O(n^{2\kappa'+\tau})\right] \geq 1 - c_2 \exp(-c_3 n^{1-2\kappa'}),$$

where c_2 and c_3 are as defined in Lemma 2.1.

Lemma 2.2 bounds the neighborhood size of the thresholded correlation network and thus the size of the conditioning sets of the ψ-partial correlation coefficient. Since the exact value of $2\kappa' + \tau$ is unknown, Liang et al. [93] suggested to bound the neighborhood size by $O(n/\log(n))$ in Algorithm 2.1. Further, Lemma 2.3 shows that correlation screening can lead to a consistent estimator for the correlation network $\mathcal{E}^{(n)}$.

Lemma 2.3 *(Lemma 4; [93]) Suppose Assumptions 2.1–2.3 hold. If $\eta_n = \frac{1}{2}c_0 n^{-\kappa}$, then*

$$P[\hat{\mathcal{E}}_{\eta_n}^{(n)} = \mathcal{E}^{(n)}] = 1 - o(1), \quad \text{as } n \to \infty.$$

By Lemmas 2.1–2.3, we have

$$P[E^{(n)} \subseteq \hat{\mathcal{E}}_{\eta_n}^{(n)}] = 1 - o(1), \tag{2.21}$$

which, together with Lemmas 2.1 and 2.2, suggests to restrict the neighborhood size of each node to

$$\min\left\{|\hat{\mathcal{E}}_{\eta_n,i}|, \frac{n}{\xi_n \log(n)}\right\}. \tag{2.22}$$

where η_n is determined in Algorithm 2.1 through multiple hypothesis tests for the correlation coefficients. Finally, Theorem 2.2 concludes the consistency of ψ-learning for structure recovery of the Gaussian graphical model.

Theorem 2.2 *(Theorem 2; [93]) Consider a Gaussian graphical model. If Assumptions 2.1–2.5 hold, then*

$$P[\hat{E}_{\zeta_n}^{(n)} = E^{(n)}] \geq 1 - o(1), \quad \text{as } n \to \infty.$$

2.8   Parallel Implementation

Before talking about the parallel implementation of the ψ-learning algorithm, let's compare its computational complexity to the other methods. The computational complexity of ψ-learning is of $O(p^2(\log p)^m)$, where $m = 3(2\kappa' + \tau)/\delta$, $O((\log p)^m)$ is the complexity for inverting a neighborhood covariance matrix of size $O(n^{2\kappa'+\tau})$ by noting $p = O(\exp(n^\delta))$, and $O(p^2)$ is the total number of ψ-scores to compute. Under the high-dimensional scenario that $\delta > 0$ in Assumption 2.2, the complexity of ψ-learning is nearly $O(p^2)$.

Under the high-dimensional scenario, the nodewise regression has a computational complexity of $O(p^2(\log p)^{2/\delta})$, which is nearly the same as that of the ψ-learning method. The graphical Lasso has a computational complexity of $O(p^3)$, but an accelerated implementation [112, 181] can reduce its computational complexity to $O(p^{2+v})$ for some $0 < v \leq 1$.

Although the ψ-learning method has a comparable computational complexity with other methods, it can be easily paralleled. In particular, both correlation coefficients and ψ-partial correlation coefficients can be calculated in parallel. In addition, both the correlation screening and ψ-screening steps are done with the empirical Bayesian method [95], where a stochastic gradient descent method can be used in fitting the mixture model involved and thus it is scalable with respect to the network size. In summary, the ψ-learning method can be very efficient for large-scale datasets.

2.9   Some Remarks

Correlation screening plays a crucial role in the development of the ψ-learning method, which reduces the dimension of the conditioning set in the evaluation of partial correlation coefficients. Although the sure screening property holds for correlation screening as shown in Lemma 2.1, exceptional cases do exist. For example, consider the case that X_i and X_j are independent and $X_k = aX_i + bX_j + \varepsilon$, where a and b are nonzero constants a and b, and ε is Gaussian random error. In this case, $X_i \not\perp\!\!\!\perp X_j | X_k$ holds, while X_i and X_j are marginally independent; therefore, $j \notin \hat{\mathcal{E}}_{\gamma_n,i}^{(n)}$ and $i \notin \hat{\mathcal{E}}_{\gamma_n,j}^{(n)}$ might hold, and the condition $n_G(i) \subseteq \hat{\mathcal{E}}_{\gamma_n,i}^{(n)}$ might be violated. As a result, the Gaussian graphical model learnt by the ψ-learning method might be slightly denser than the true one, since some ψ-partial correlation coefficients are calculated based on an incomplete conditioning set. This flaw of marginal correlation screening can be easily fixed by including a neighborhood amendment step after correlation screening, i.e., for each node i, finding $\Delta_i = \{ j : j \notin \hat{\mathcal{E}}_{\gamma_n,i}^{(n)}, \exists k \in \hat{\mathcal{E}}_{\gamma_n,i}^{(n)} \cap \hat{\mathcal{E}}_{\gamma_n,j}^{(n)} \}$ and updating the neighboring set by $\hat{\mathcal{E}}_{\gamma_n,i}^{(n)} \leftarrow \hat{\mathcal{E}}_{\gamma_n,i}^{(n)} \cup \Delta_i$. However, in practice, this issue is usually not a big concern, as we often expect $E^{(n)} \subseteq \hat{E}_{\zeta_n}^{(n)}$ to hold with a finite sample size of n.

2.10   Problems

1. Prove the problem (2.2) is convex.

2. Prove Lemma 2.1.

3. Prove Lemma 2.2.

4. Prove Lemma 2.3.

5. Prove Theorem 2.2.

6. Liang et al. [93] pointed out that the ψ-learning method suffers from the phase transition phenomenon of partial correlation screening, i.e., it tends to outperform other methods in the low-recall region, but not in the high-recall region. Redo the simulation example in Section 2.4 to observe this phenomenon by plotting the precision-recall curves.

7. Simulate two datasets from the model (2.10) to test the data integration method described in Section 2.6.1.

8. Simulate two datasets from the model (2.10) to test the network comparison method described in Section 2.6.2.

9. Simulate some datasets to test the covariate adjustment method described in Section 2.6.3.

10. Modify Algorithm 2.1 by replacing step (a) by nodewise regression with small regularization parameter values and redo the simulation example with the modified algorithm.

Chapter 3

Gaussian Graphical Modeling with Missing Data

This chapter describes how to use the ψ-learning method under the framework of the imputation-regularized optimization (IRO) algorithm [92] to learn high-dimensional Gaussian graphical models when missing data are present. The IRO algorithm is critical for high-dimensional missing data problems. Through imputation, it converts the missing data problem to a pseudo-complete data problem such that many of the existing high-dimensional statistical methods and theories can be applied. The IRO algorithm can significantly enrich the applications of high-dimensional statistics in modern data science.

3.1 Introduction

Missing data are ubiquitous throughout many fields of modern data science, such as biomedical science and social science. When missing data are present, one simple way is to exclude them from analysis by deleting the corresponding samples or variables from the dataset. Obviously, this can cause a significant loss of data information when the missing rate of the dataset is high, or even a biased estimate when missing is not at random. Alternatively, one might fill each missing value by the mean or median of the corresponding variable, but this will potentially lead to a biased inference for the model as the uncertainty of the missing values is not properly accounted for in the filling. The same issue can also happen to other one-time imputation methods such as those based on matrix decomposition [125, 166, 171] and least square methods [15]. To address this issue, the Expectation-Maximization (EM) [35] and multiple imputation [91] types of algorithms have been developed for missing data problems under various contexts. In particular, for high-dimensional Gaussian graphical models, the MissGLasso [155] and imputation-regularized optimization (IRO) [92] algorithms have been developed, which will be briefly described in the following sections of this chapter.

3.2 The MissGLasso Algorithm

Consider a set of Gaussian random variables $(X_1, X_2, \ldots, X_p) \sim N(\mu, \Sigma)$. Let $x :=$ $(x_{\text{obs}}, x_{\text{mis}})$ denote complete data, where x_{obs} denotes the set of observed data and

DOI: 10.1201/9780429061189-3

x_{mis} denotes missing data. Let $x^{(i)} = (x_{\text{obs}}^{(i)}, x_{\text{mis}}^{(i)})$ denote the ith observation in x. Let $\Theta = \Sigma^{-1}$ denote the inverse of the covariance matrix Σ.

The MissGLasso algorithm [155] is an extension of the graphical Lasso algorithm [49, 193] to the case with missing data in presence, which estimates μ and Θ by solving the minimization problem:

$$\hat{\mu}, \hat{\Theta} = \arg \min_{(\mu, \Theta):\Theta > 0} \left\{ -l(\mu, \Theta; x_{\text{obs}}) + \lambda \|vec(\Theta)\|_1 \right\}, \tag{3.1}$$

where λ is a regularization parameter, the log-likelihood function of the observed data is given by

$$l(\mu, \Theta; x_{\text{obs}}) = -\frac{1}{2} \sum_{i=1}^{n} \left(\log |\Sigma_{\text{obs}}^{(i)}| + (x_{\text{obs}}^{(i)} - \mu_{\text{obs}}^{(i)})^T (\Sigma_{\text{obs}}^{(i)})^{-1} (x_{\text{obs}}^{(i)} - \mu_{\text{obs}}^{(i)}) \right),$$

and $\mu_{\text{obs}}^{(i)}$ and $\Sigma_{\text{obs}}^{(i)}$ denote, respectively, the mean and covariance matrix of the observed component of x for observation i. The minimization problem in equation (3.1) can be solved using the EM algorithm based on the property of the multivariate Gaussian distribution.

3.2.1   E-Step

For the multivariate Gaussian distribution, when the data are fully observed, the log-likelihood function can be expressed as a linear function of the sufficient statistics $(T_1, T_2) := (\sum_{i=1}^{n} x^{(i)}, \sum_{i=1}^{n} x^{(i)} (x^{(i)})^T)$, i.e.,

$$l(\mu, \Theta; x) = \frac{n}{2} \log |\Theta| - \frac{n}{2} \mu^T \Theta \mu + \mu^T \Theta T_1 - \frac{1}{2} tr(\Theta T_2).$$

Based on this mathematical fact, the E-step of the algorithm proceeds to calculate the Q-function

$$Q(\mu, \Theta | \mu_t, \Theta_t) = \mathbb{E}[l(\mu, \Theta; x) | \mu_t, \Theta_t, x_{\text{obs}}] - \lambda \|vec(\Theta)\|_1,$$

through calculating the conditional expectations:

$$T_{1,t+1} = \mathbb{E}[T_1 | \mu_t, \Theta_t, x_{\text{obs}}],$$
$$T_{2,t+1} = \mathbb{E}[T_2 | \mu_t, \Theta_t, x_{\text{obs}}],$$

where μ_t and Θ_t denote, respectively, the estimates of μ and Θ at iteration t. Explicit formulas for calculating $T_{1,t+1}$ and $T_{2,t+1}$ can be derived based on the conditional mean function of the multivariate Gaussian distribution. In particular, evaluation of $T_{1,t+1}$ involves the formula

$$\mathbb{E}[x_{ij} | x_{\text{obs}}^{(i)}, \mu_t, \Theta_t] = \begin{cases} x_{ij}, & \text{if } x_{ij} \text{ is observed,} \\ c_{ij}, & \text{if } x_{ij} \text{ is missed,} \end{cases} \tag{3.2}$$

where x_{ij} denotes the jth component of $x^{(i)}$, and c_{ij} is a component of the conditional mean vector

$$c_i := \mu_{mis,t} - (\Theta^{(t)}_{mis,mis})^{-1}\Theta^{(t)}_{mis,obs}(x^{(i)}_{obs} - \mu_{obs,t}), \qquad (3.3)$$

$\Theta^{(t)}_{mis,mis}$ is the sub-matrix of Θ_t with rows and columns corresponding to the missing variables for case i, and $\Theta^{(t)}_{mis,obs}$ is the sub-matrix of Θ_t with rows corresponding to the missing variables and columns corresponding to observed variables for case i. It is obvious that the calculation of the inverse matrix $(\Theta^{(t)}_{mis,mis})^{-1}$ can be time-consuming when the number of missing variables in case i is large. Refer to [155] for the formulas for the evaluation of $T_{2,t+1}$.

3.2.2 M-Step

It calculates $(\mu_{t+1}, \Theta_{t+1})$ as the maximizer of the Q-function $Q(\mu, \Theta | \mu_t, \Theta_t)$ by setting

$$\mu_{t+1} = \frac{1}{n} T_{1,t+1}, \qquad (3.4)$$

$$\Theta_{t+1} = \arg\max_{\Theta > 0} \left(\log |\Theta| - tr(\Theta\widehat{\Sigma}_{t+1}) - \frac{2\lambda}{n} \|vec(\Theta)\|_1 \right), \qquad (3.5)$$

where $\widehat{\Sigma}_{t+1} = \frac{1}{n} T_{2,t+1} - \mu_{t+1}\mu_{t+1}^T$, and Θ_{t+1} is calculated as in Glasso through the coordinate ascent algorithm [169].

Städler and Bühlmann [155] showed that the algorithm will converge to a stationary point of the penalized likelihood function in equation (3.1).

3.3 The Imputation-Regularized Optimization Algorithm

Unlike the MissGLasso algorithm, which is specially designed for learning high-dimensional Gaussian graphical models when missing data are present, the IRO algorithm [92] is designed for general high-dimensional missing data problems. Let θ denote the parameter vector of the model under study. The IRO algorithms start with an initial guess θ_0 and then iterates between the following two steps:

Algorithm 3.1 *(IRO Algorithm; [92])*
- **Imputation** *(I-step): Draw $x_{mis,t+1}$ from the predictive distribution $h(x_{mis}|x_{obs}, \theta_t)$ given x_{obs} and θ_t.*

- **Regularized optimization** *(RO-step): On the basis of the pseudo-complete data $(x_{obs}, x_{mis,t+1})$, calculate θ_{t+1} which forms a consistent estimate of*

$$\tilde{\theta}_{t+1} = \arg\max_{\theta} \mathbb{E}_{\theta_t}[\log f_\theta(x_{obs}, x_{mis,t+1})], \qquad (3.6)$$

where $\mathbb{E}_{\theta_t}[\log f_\theta(x_{obs}, x_{mis,t+1})] = \int \log f(x_{obs}, x_{mis,t+1}; \theta) f(x_{obs}; \theta^*) h(x_{mis,t+1}| x_{obs}, \theta_t)dx_{obs}dx_{mis,t+1}$, θ^* denotes the true value of the parameters, and $f(x_{obs}; \theta^*)$ denotes the marginal density function of x_{obs}.

As in the stochastic EM algorithm [20, 123], the IRO algorithm leads to two interleaved Markov chains:

$$\theta_0 \rightarrow x_{\text{mis},1} \rightarrow \theta_1 \rightarrow x_{\text{mis},2} \rightarrow \cdots \rightarrow \theta_t \rightarrow x_{\text{mis},t+1} \rightarrow \cdots. \tag{3.7}$$

Under general conditions, Liang et al. [92] showed that $\|\theta_t - \theta^*\| \xrightarrow{p} 0$ as the sample size $n \rightarrow \infty$ and the iteration number $t \rightarrow \infty$, where the dependency of θ_t on the sample size has been depressed for simplicity of notation. The key to the convergence of the IRO algorithm is to find a uniformly consistent (over iterations) sparse estimator for the working "true" parameter $\tilde{\theta}_t$ under the high-dimensional scenario. As shown by Liang et al. [92], such an estimator can typically be obtained in two ways. One way is to maximize an appropriately penalized log-likelihood function, and the other way is to first restrict the space of $\tilde{\theta}_t$ to some low-dimensional subspace through a sure screening procedure and then find a consistent estimator, e.g., maximum likelihood estimator, in the low-dimensional subspace. Also, Liang et al. [92] pointed out that the sure screening procedure can be viewed as a special regularization method, which imposes a zero penalty on the solutions in the low-dimensional subspace and an infinitely large penalty on those outside the low-dimensional subspace.

The IRO algorithm can be easily used for learning Gaussian graphical models when missing data are present. For example, the ψ-learning method can be used as a consistent estimator required in the RO-step. As shown in Chapter 2, the ψ-learning method belongs to the class of sure screening methods and provides a consistent estimate for the structure of the Gaussian graph. In this context, θ, with a slight abuse of notation, can be understood as the concentration matrix of the Gaussian graphical model, which can be uniquely determined from the network structure with a modified regression algorithm ([60], p. 634).

To detail the algorithm, we let $E_t = (e_{ij}^{(t)})$ denote the adjacency matrix of the Gaussian graph learned at iteration t. Let x_{ij} denote a missing entry in $x^{(i)}$. Let $\omega_t(j) = \{k : e_{jk}^{(t)} = 1, k = 1, 2, \ldots, p\}$ denote the neighboring set of node j in the current network E_t, which forms the Markov blanket of node j by the Markov property of Gaussian graphical models. Therefore, the variable X_j is independent of all other variables conditioned on $X_{j\omega_t} := \{X_k : k \in \omega_t(j)\}$, and x_{ij} can be imputed conditioned on $x_{j\omega_t}^{(i)}$ only, where $x_{j\omega_t}^{(i)}$ denotes the realizations of $X_{j\omega_t}$ in observation i. For the time being, we can assume that $x_{j\omega_t}^{(i)}$ contains no missing values. Mathematically, we have

$$\begin{pmatrix} X_j \\ X_{j\omega_t} \end{pmatrix} \sim N \left(\begin{pmatrix} \mu_j \\ \mu_{\omega_t} \end{pmatrix}, \begin{pmatrix} \sigma_j^2 & \Sigma_{j\omega_t} \\ \Sigma_{j\omega_t}^T & \Sigma_{\omega_t\omega_t} \end{pmatrix} \right), \tag{3.8}$$

where $\mu_j, \mu_{\omega_t}, \sigma_j^2, \Sigma_{j\omega_t}$ and $\Sigma_{\omega_t\omega_t}$ denote the corresponding mean and variance components. The conditional distribution of X_j given $X_{j\omega_t}$ is Gaussian with the mean and variance given by

$$\mu_{j|\omega_t} = \mu_j + \Sigma_{j\omega_t} \Sigma_{\omega_t\omega_t}^{-1} (X_{j\omega_t} - \mu_{\omega_t}), \quad \sigma_{j|\omega_t}^2 = \sigma_j^2 - \Sigma_{j\omega_t} \Sigma_{\omega_t\omega_t}^{-1} \Sigma_{j\omega_t}^T. \tag{3.9}$$

By the ψ-learning method [93], the neighboring set size of each node can be upper bounded by $\lceil n/\log(n) \rceil$, where $\lceil z \rceil$ denotes the smallest integer not smaller than z.

Therefore, $\Sigma_{\omega_t \omega_t}$ can be estimated by the standard empirical sample variance $S_{\omega_t \omega_t}$, whose nonsingularity is ensured by the bounded neighboring set size. Let s_j^2, $S_{j\omega_t}$, $\bar{x}_j$ and $\bar{x}_{\omega_t}$ denote the respective empirical estimates of σ_j^2, $\Sigma_{j\omega_t}$, μ_j and μ_{ω_t}. Then x_{ij} can be imputed by drawing from the Gaussian distribution

$$N(\bar{x}_j + S_{j\omega_t} S_{\omega_t \omega_t}^{-1} (x_{i\omega_t} - \bar{x}_{\omega_t}), s_j^2 - S_{j\omega_t} S_{\omega_t \omega_t}^{-1} S_{j\omega_t}^T). \tag{3.10}$$

To impute all missing entries, we can run the Gibbs sampler [56] for one or a few iterations by iteratively drawing from the corresponding conditional distributions as given in equation (3.10).

Compared to the imputation scheme of missGLasso, it is easy to see that the imputation scheme of the IRO algorithm is potentially more efficient. The former involves inverting the matrix $\Theta_{\text{mis,mis}}^{(t)}$, see equation (3.3), which can be more time-consuming when the number of missing entries is large. In contrast, by the Markov property of Gaussian graphical models, the IRO algorithm has successfully restricted the corresponding computation to the matrices of neighboring set size.

In summary, we have the following algorithm for the reconstruction of Gaussian graphical models when missing data are present:

Algorithm 3.2 *(IRO-Ave algorithm; [92])*

- *(Initialization) Fill each missing entry by the median of the corresponding variable, and then iterates between the RO- and I-steps.*

- *(RO-step) Apply the ψ-learning method to learn the network structure of a Gaussian graphical model.*

- *(I-step) Based on the network structure learnt in the RO-step, impute the missing entries by running the Gibbs sampler for one or a few iterations with the conditional distributions as given in equation (3.10).*

The IRO algorithm generates a series of Gaussian graphical networks along with iterations. To integrate these networks into a single one, the ψ-integration method as described in Section 2.6.1 can be used. Given the averaged ψ-scores, the network structure can be identified through a multiple hypothesis test [95], see Appendix A.7 for a brief description of the method.

3.4   Simulation Studies

In this study, we generated the datasets from a Gaussian graphical model with an AR(2) structure as defined in Section 2.4. We fixed the sample size $n = 200$ and varied the dimension $p = 100, 200$, and 300. Under each setting of (n, p), we generated ten datasets independently and then randomly delete 10% of the observations as missing values. The IRO algorithm was run for each dataset for 30 iterations, where the first 10 iterations were discarded as the burn-in period, and the ψ-partial correlation coefficients generated in the other iterations were used for inference of the network structure. For comparison, the missGLasso algorithm[1] was also applied to the same datasets.

---

[1]It is implemented in the R package *spaceExt* [63].

The results are summarized in Figure 3.1 and Table 3.1. The former compares the precision-recall curves produced by the two methods for three datasets of different dimensions, and the latter compares the averaged AUC values (over 10 datasets) of the precision-recall curves. The comparison indicates the superiority of the IRO-Ave

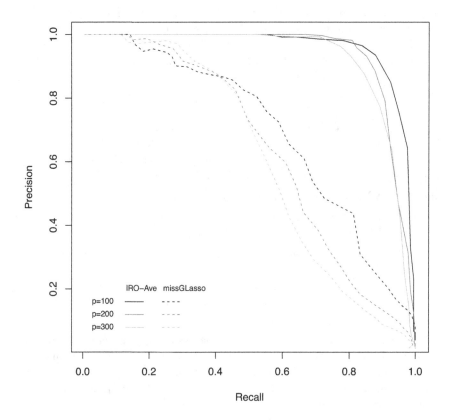

Figure 3.1 *Precision-recall curves produced by IRO-Ave and missGLasoo for three datasets of different dimensions.*

Table 3.1 *Average AUC values (over ten datasets) under the precision-recall curves produced by IRO-Ave and missGLasso, where the number in the parentheses represents the standard deviation of the average AUC value.*

p	missGLasso	IRO-Ave
100	0.6698 (0.0049)	0.9424 (0.0029)
200	0.6226 (0.0029)	0.9283 (0.0021)
300	0.5931 (0.0037)	0.9171 (0.0028)

method over missGLasso. In Liang et al. [92], more methods such as median filling, Bayesian principal component analysis[2] (BPCA) [125], and multiple imputation regression tree (MIRegTree)[3] have been included for comparison. Refer to [92] for the detail.

3.5 Application

Gasch et al. [55] studied how the gene expression levels of the yeast *Saccharomyces cerevisiae* change in response to environmental changes such as temperature shocks and nitrogen source depletion, among others. The dataset[4] consists of 173 samples and 6,152 genes, and it has been analyzed by Liang et al. [92] using the IRO algorithm. Their results are briefly reviewed as follows.

In the analysis, Liang et al. [92] considered only top 1,000 genes with the highest variation in expression levels over the samples. The missing rate for the sub-dataset is 3.01%. They ran the IRO algorithm for the sub-dataset for 30 iterations, and they reconstructed the gene regulatory network by integrating those obtained over the last 10 iterations. Based on the reconstructed gene regulatory network, they identified three hub genes, ARB1, HXT5, and XBP1, which are all included in the transcriptional factor (TF)-gene regulatory network[5] [188]. Note that YTRP deposits TF-gene regulatory pairs inferred from TF-perturbation experiments, which can somehow be considered as "ground truth". Also, they found that the gene regulatory network learned by the IRO-Ave algorithm obeys the power law (see Appendix A.3). For comparison, Liang et al. [92] have tried other methods such as missGLasso, median filling, BPCA, and MIRegTree on this example, but the results of these methods are less favorable than that by IRO-Ave. Refer to [92] for detail.

3.6 Problems

1. Prove the chain $\{\theta_t : t = 0,1,\ldots\}$ induced by the IRO algorithm as given in equation (3.7) is ergodic.

2. Prove that Algorithm 3.2 leads to a consistent estimator for the network structure of the Gaussian graphical model when missing data are present.

3. Redo the simulation example in Section 3.4 by replacing the ψ-learning method in the RO-step of Algorithm 3.2 by the nodewise Lasso regression method.

4. Redo the simulation example in Section 3.4 by replacing the ψ-learning method in the RO-step of Algorithm 3.2 by the nodewise SCAD regression method.

5. Redo the simulation example in Section 3.4 by replacing the ψ-learning method in the RO-step of Algorithm 3.2 by the nodewise MCP regression method.

---

[2]It has been implemented in the package *pcaMethods* [154].
[3]it has been implemented in the R package *MICE* [172].
[4]The dataset is available at http://genome-www.stanford.edu/yeast-stress/.
[5]The network is available at http://cosbi3.ee.ncku.edu.tw/YTRP.

Chapter 4

Gaussian Graphical Modeling for Heterogeneous Data

This chapter introduces the mixture ψ-learning method [73] for learning Gaussian graphical models with heterogeneous data, which provides a mechanism for integrating the structure information from different clusters hidden in the data. Our numerical experience shows that the data heterogeneity can significantly affect the performance of Gaussian graphical modeling methods, e.g., nodewise regression, graphical Lasso, and ψ-learning, which are developed under the data homogeneity assumption. A data homogeneity test seems necessary before the application of these methods.

4.1 Introduction

In previous chapters, we have studied how to construct Gaussian graphical models under the assumption that the data are homogeneous, i.e., all samples are from the same multivariate Gaussian distribution, although missing data are allowed to be present. However, due to the varied nature of our world, the data are often heterogeneous, i.e., the samples can be drawn from a mixture Gaussian distribution. The gene expression data can be heterogeneous due to many reasons such as existence of unknown disease subtypes or systematic differences in sample collection (e.g., operated by different people and/or at different time). For such types of heterogeneous data, we are still interested in learning a single Gaussian graphical model, as the underlying variable regulatory mechanisms might be highly similar for different components of the mixture distribution. Moreover, it might be impractical to learn a different Gaussian graphical model for each component separately, as the number of samples from each component of the mixture distribution can be small. This problem has been addressed by Jia and Liang [73], who proposed a method for jointly clustering the samples and learning the graphical model. Their method is described in this chapter.

The problem on how to deal with data heterogeneity in Gaussian graphical modeling has also been considered by other authors in the literature. For example, Ruan et al. [140] proposed a mixture graphical Lasso method, where the mixture proportion and mean of each component are learned as in the standard EM algorithm [35], while the concentration matrix of each component is learned through the graphical

DOI: 10.1201/9780429061189-4

Lasso algorithm [49, 193]. Another related work is by Lee et al. [89], where the high-dimensional samples are first clustered using a singular value decomposition-based method and then a single Gaussian graphical model is learned using the ψ-integration method (see Section 2.6.1) by combining the data information from all clusters.

4.2 Mixture Gaussian Graphical Modeling

Let $\mathcal{X} = \{x^{(1)}, \ldots, x^{(n)}\}$ denote a set of n independent p-dimensional samples drawn from a mixture Gaussian distribution,

$$\pi_1 \mathcal{N}_p(\mu_1, \Sigma_1) + \pi_2 \mathcal{N}_p(\mu_2, \Sigma_2) + \cdots + \pi_M \mathcal{N}_p(\mu_M, \Sigma_M), \qquad (4.1)$$

where $\mathcal{N}_p(\mu, \Sigma)$ denotes a p-dimensional multivariate Gaussian distribution with mean μ and covariance matrix Σ, π_k's are mixture proportions such that $0 \leq \pi_1, \pi_2, \ldots, \pi_M \leq 1$ and $\sum_{i=1}^{M} \pi_k = 1$, and M is the number of components of the mixture distribution. For the time being, let's assume that M is known. Let $\theta = \{\pi_1, \pi_2, \ldots, \pi_{M-1}; \mu_1, \mu_2, \ldots, \mu_M; \Sigma_1, \Sigma_2, \ldots, \Sigma_M\}$ denote the set of unknown parameters of the model. Let $\tau = (\tau_1, \tau_2, \ldots, \tau_n)$ give the cluster memberships of the n samples, where $\tau_i \in \{1, 2, \ldots, M\}$. Since our goal is to learn a single Gaussian graphical model by integrating the data from all clusters, we further make the following assumption:

Assumption 4.1 *All components of the mixture Gaussian distribution (equation 4.1) share a common network structure, while their covariance/concentration matrices can be different from each other.*

By viewing τ as missing data, the parameter θ can then be estimated using the imputation-regularized optimization (IRO) algorithm [92] even under the high-dimensional scenario, i.e., $n < p$. For the IRO algorithm, the common network structure can be estimated using the ψ-integration method as described in Section 2.6.1, which is consistent as implied by the theory developed by Liang et al. [93]. Given the common network structure, the covariance matrix of each component can be uniquely recovered by the modified regression algorithm given in Hastie et al. [60] (p. 634). Further, the mixture proportions and component means can be consistently estimated based on the samples assigned to each cluster. In summary, we have the following algorithm for learning Gaussian graphical models with heterogeneous data. Let $\theta^{(t)} = \{\pi_1^{(t)}, \pi_2^{(t)}, \ldots, \pi_{M-1}^{(t)}; \mu_1^{(t)}, \mu_2^{(t)}, \ldots, \mu_M^{(t)}; \Sigma_1^{(t)}, \Sigma_2^{(t)}, \ldots, \Sigma_M^{(t)}\}$ denote the estimate obtained at iteration t. The algorithm starts with an initial estimate $\theta^{(0)}$ and then iterates in the following steps:

Algorithm 4.1 *(Mixture ψ-learning method; [73])*

- *(Imputation) For each sample i, impute the cluster memberships $\tau_i^{(t+1)}$ by drawing from the multinomial distribution*

$$P(\tau_i = k | x^{(i)}, \theta^{(t)}) = \frac{\pi_k^{(t)} \phi(x^{(i)}; \mu_k^{(t)}, \Sigma_k^{(t)})}{\sum_{l=1}^{M} \pi_l^{(t)} \phi(x_i^{(t)}; \mu_l^{(t)}, \Sigma_l^{(t)})}, \quad k = 1, 2, \ldots, M, \qquad (4.2)$$

where $\pi(x;\mu,\Sigma)$ denotes the density function of a multivariate Gaussian distribution with mean μ and covariance matrix Σ.

- *(Regularized Optimization) Conditioned on the cluster memberships $\{\tau_i^{(t+1)} : i = 1,2,\ldots,n\}$, update the estimate $\theta^{(t)}$ as follows:*

    i. *For each cluster $k \in \{1,2,\ldots,M\}$, set $n_k^{(t+1)} = \sum_{i=1}^n I(\tau_i^{(t+1)} = k)$, $\pi_k^{(t+1)} = n_k^{(t+1)}/n$, and $\mu_k^{(t+1)} = \sum_{j \in \{i:\tau_i^{(t+1)}=k\}} x^{(j)}/n_k^{(t+1)}$.*

    ii. *For each cluster, apply the ψ-learning method to learn a network structure.*

    iii. *Apply the ψ-integration method (see Section 2.6.1) to integrate the networks learned for different clusters into a single network.*

    iv. *Applying the modified regression algorithm to recover the covariance matrix of each cluster, given the common network learned in (iii).*

As mentioned in Section 3.3, the IRO algorithm generates two interleaved Markov chains:

$$\theta^{(0)} \to \tau^{(1)} \to \theta^{(1)} \to \ldots \to \tau^{(t)} \to \theta^{(t)} \to \cdots$$

By assuming that the sample size is sufficiently large and some other mild conditions hold, Liang et al. [92] showed that the Markov chain $\{\theta^{(t)} : t = 0,1,\ldots\}$ is ergodic almost surely, and the path average of $\theta^{(t)}$'s forms a consistent estimator for the true parameter θ^* as t becomes large. To comply with this theory, Jia and Liang [73] further averaged the network structures obtained at each iteration of Algorithm 4.1 through the ψ-integration method.

The substeps (iii) and (iv) in Algorithm 4.1 can be implemented as follows. Let $\psi_{kij}^{(t)}$ denote the ψ-partial correlation coefficient of variable i and variable j in cluster k at iteration t, respectively. It can be transformed to a z-score by Fisher's z-transformation:

$$Z_{kij}^{(t)} = \frac{\sqrt{n_k^{(t)} - |S_{kij}^{(t)}| - 3}}{2} \log \left[\frac{1 + \psi_{kij}^{(t)}}{1 - \psi_{kij}^{(t)}} \right], \quad i,j = 1,\ldots p, k = 1,\ldots M. \quad (4.3)$$

where $S_{kij}^{(t)}$ denotes the conditioning set used in the calculation of $\psi_{kij}^{(t)}$. The z-scores from different clusters can be integrated via Stouffer's meta-analysis method [157],

$$Z_{ij}^{(t)} = \frac{\sum_{k=1}^M \omega_k^{(t)} z_{kij}^{(t)}}{\sqrt{\sum_{k=1}^M (\omega_k^{(t)})^2}}, \quad i,j = 1,\ldots p, \quad (4.4)$$

where $\omega_k^{(t)} = n_k^{(t)}/n$ is the weight assigned to cluster k at iteration t. A multiple hypothesis test on $Z_{ij}^{(t)}$ completes the substep (iii). The integrated z-scores can be further integrated over iterations of Algorithm 4.1 by setting

$$Z_{ij} = \frac{1}{T - t_0} \sum_{t=t_0+1}^T Z_{ij}^{(t)}, \quad i,j = 1,2,\ldots,p,$$

where t_0 is the burn-in period of the algorithm, and Z_{ij} can be compared to the standard Gaussian distribution for the test. Note that as the sample size $n \to \infty$ and the iteration number $t \to \infty$, $\theta^{(t)} \xrightarrow{P} \theta^*$ holds and the network structure can be consistently identified along with iterations and, therefore, $Z_{ij}^{(t)}$ converges to a constant score in probability. The final estimate for the network structure can be obtained via a multiple hypothesis test on Z_{ij}'s. Again, Jia and Liang [73] employed the empirical Bayesian method by Liang and Zhang [95] to perform the multiple hypothesis tests involved in the method.

Jia and Liang [73] suggested to determine the number of clusters M by minimizing the BIC value. By following Ruan et al. [140], they defined the degree of freedom for the model (4.1) as

$$df(M) = M\left(p + \sum_{i \leqslant j} e_{ij}\right), \tag{4.5}$$

where p is the dimension of the mean vector, e_{ij} is the $(i, j)^{th}$ element of adjacency matrix of the common network structure, and $\sum_{i \leqslant j} e_{ij}$ represents the number of parameters contained in the concentration matrix of each component of the model. Recall Assumption 4.1, we have assumed that all components share a common network structure while having different concentration matrices. The BIC value for the model (4.1) is then given by

$$BIC(M) = -2\ell(\mathcal{X}|\theta) + \log(n)df(M), \tag{4.6}$$

where $\ell(\mathcal{X}|\theta)$ denotes the log-likelihood function of the model.

4.3   Simulation Studies

The datasets were simulated from a three-component mixture Gaussian distribution with the respective concentration matrix $\Theta^{(k)} = (\theta_{ij}^{(k)})$ given by

$$\theta_{ij}^{(k)} = \begin{cases} 1, & i = j, \\ c_k, & |j-i| = 1, \\ c_k/2, & |j-i| = 2, \\ 0, & \text{otherwise,} \end{cases} \tag{4.7}$$

for $k = 1, 2, 3$, where $c_1 = 0.6$, $c_2 = 0.5$ and $c_3 = 0.4$. This setting of $\Theta^{(k)}$'s complies with Assumption 4.1 that all components share a common network structure while their concentration matrices are different. For the mean vectors, we considered two cases. In the first case, we set $\mu_2 = \mathbf{0}_p$, and draw each component of μ_1 and μ_3 from the uniform distribution Uniform$[-1, 1]$ independently. In the second case, we set $\mu_1 = 0.51_p$, $\mu_2 = \mathbf{0}_p$, and $\mu_3 = -0.51_p$, where $\mathbf{0}_p$ and $\mathbf{1}_p$ denote a p-vector of zeros and ones, respectively. In this study, we fix the sample size $n = 300$ with 100 independent samples generated from each component and vary the dimension $p \in \{100, 200\}$. Under each setting of the dimension and mean vectors, we simulated ten datasets independently.

Table 4.1 *Average AUC values (over ten datasets) produced by nodewise regression, graphical Lasso (gLasso), ψ-learning, and mixture ψ-learning for the synthetic example, where the number in the parentheses represents the standard deviation of the average.*

p	Mean	Nodewise	gLasso	ψ-learning	Mixture ψ-learning
100	Random	0.7404	0.6403	0.9059	0.9390
		(0.0085)	(0.0080)	(0.0076)	(0.0025)
	Fixed	0.2520	0.2376	0.7224	0.9330
		(0.0051)	(0.0055)	(0.0049)	(0.0051)
200	Random	0.7166	0.6086	0.8823	0.9036
		(0.0031)	(0.0036)	(0.0045)	(0.0040)
	Fixed	0.2265	0.2125	0.6921	0.9081
		(0.0015)	(0.0016)	(0.0034)	(0.0053)

Table 4.1 compares the average AUC values (over ten datasets) for the precision-recall curves produced by nodewise regression [118], graphical Lasso [49], ψ-learning [93], and mixture ψ-learning methods. The comparison indicates that the mixture ψ-learning method significantly outperforms the other methods especially when the mean vectors of the different components of the mixture distribution are more well separated. Also, the comparison shows that the data heterogeneity can significantly affect the performance of the methods, graphical Lasso, nodewise regression, and ψ-learning, which are developed under the data homogeneity assumption.

4.4   Application

The dataset was downloaded from The Cancer Genome Atlas (TCGA)[1], which consisted of 768 subjects. For each subject, the dataset consisted of expression levels of 20,502 genes and some clinical information such as gender, age, tumor stage, and survival time. With this dataset, Jia and Liang [73] learned a regulatory network for the genes related to the survival time of breast cancer patients. Their results are briefly reviewed as follows.

In preprocessing the data, Jia and Liang [73] employed a sure independence screening method [46] to select the genes that are possibly related to the survival time. More precisely, they first calculated the p-value of each gene with a marginal Cox regression after adjusting the effects of the covariates such as age, gender, and tumor stages, and then they applied the multiple hypothesis test [95] to select the genes whose p-values are smaller than others (at a significance level of $\alpha = 0.05$). After this screening process, 592 genes were left.

They tested the heterogeneity of the reduced dataset by fitting it with a mixture Gaussian model. They found that the data are heterogeneous, most likely consisting of three clusters according to the BIC values. By setting $M = 3$, the mixture

---

[1]The data can be accessed at http://cancergenome.nih.gov/.

ψ-learning method divided the subjects into three clusters with the respective cluster size 338, 191, and 238. They compared the Kaplan-Meier curves of the three groups of subjects. A log-rank test for the comparison produced a p-value of 3.89×10^{-5}, which indicates that the three groups of subjects might have significantly different survival times. They justified their clustering results by a literature review, which shows that the results might be biologically meaningful and the data heterogeneity is likely caused by the hidden subtypes of breast cancer. By Haque et al. [59], the breast cancer has four molecular subtypes, including luminal A, luminal B, basal-like, and HER2-enriched, and different subtypes have different survival time. In particular, luminal A patients have longer survival time, the luminal B and HER2-enriched patients have shorter survival time, and the basal-like patients are in the middle.

In a cross-validation study for the dataset with the mixture ψ-learning method, they identified ten hub genes and found that eight of them have been verified in the literature as breast cancer-associated genes. Refer to [73] for the detail.

4.5 Problems

1. Prove the consistency of Algorithm 4.1 for structure learning of the Gaussian graphical model when the data are heterogeneous.

2. Redo the simulation example in Section 4.3.

3. Perform a simulation study for a mixture Gaussian distribution with a common concentration matrix but different mean vectors.

4. Extend Algorithm 4.1 for suiting the purpose of learning different networks for different clusters and performing a simulation study.

5. Extend Algorithm 4.1 to the case with missing data in presence and perform a simulation study.

Chapter 5

Poisson Graphical Models

This chapter introduces a method for learning gene regulatory networks with next generation sequencing (NGS) data based on the work [74]. The method works in three steps: It first conducts a random effect model-based transformation to continuize NGS data, then transforms the continuized data to Gaussian through a semi-parametric transformation, and finally applies the ψ-learning method to recover gene regulatory networks. The data-continuized transformation has some interesting implications. In particular, the logarithms of RNA-seq data have often been analyzed as continuous data, though not rigorously, in the literature; the data-continuized transformation provides a justification for this treatment.

5.1 Introduction

Next generation sequencing (NGS) has revolutionized transcriptome studies, say, through sequencing RNA (RNA-seq). The RNA-seq quantifies gene expression levels by counts of reads, which has many advantages, such as higher throughput, higher resolution, and higher sensitivity to gene expression levels, compared to the microarray platform. However, the discrete nature of the RNA-seq data has caused many difficulties to the following statistical analysis. In particular, the RNA-seq data are traditionally modeled by Poisson [159] or negative binomial distributions [4, 138], but based on which the gene regulatory networks (GRNs)are hard to be learned due to some intrinsic difficulties as discussed below.

Consider a Poisson graphical model (PGM) $G = (V, E)$, for which each node X_j represents a Poisson random variable, and the nodewise conditional distribution is given by

$$P(X_j | X_k, \forall k \neq j; \Theta_j) = \exp\left\{ \theta_j X_j - \log(X_j!) + \sum_{k \neq j} \theta_{jk} X_j X_k - A(\Theta_j) \right\}, \quad (5.1)$$

$$j = 1, 2, \ldots, p,$$

where $\Theta_j = \{\theta_j, \theta_{jk}, k \neq j\}$ are the coefficients associated with the Poisson regression $X_j \sim \{X_k : k \neq j\}$, and $A(\Theta_j)$ is the log-partition function of the conditional distribution. By the Hammersley-Clifford theorem [14], the joint distribution of

DOI: 10.1201/9780429061189-5

$(X_1, X_2, \ldots, X_p)$ can be obtained by combining the conditional distributions in equation (5.1), which yields

$$P(X; \Theta) = \exp\left[\sum_{j=1}^{p} (\theta_j X_j - \log(X_j!)) + \sum_{j \neq k} \theta_{jk} X_j X_k - \phi(\Theta)\right], \qquad (5.2)$$

where $\Theta = (\Theta_1, \ldots, \Theta_p)$ and $\phi(\Theta)$ is the log-partition function. However, to ensure $\phi(\Theta)$ to be finite, the interaction parameters θ_{jk} must be non-positive for all $j \neq k$ [14, 185]. That is, the PGM permits negative conditional dependencies only, which makes the method of Poisson graphical modeling impractical. By Patil and Joshi [127], the joint negative binomial distribution also suffers from the same issue.

To address this issue, local model-based methods have been developed, see e.g. [3, 53, 70, 187], which ignore the joint distribution (equation 5.2) and works in a similar way to nodewise regression [118]. More precisely, it first fits a sparse Poisson regression model for each gene with a regularization method according to equation (5.1), and then union the sparse Poisson regression models to form a graphical model. However, due to ignorance of the joint distribution (equation 5.2), the conditional dependence $X_k \perp\!\!\!\perp X_j | X_{V \setminus \{k,j\}}$ is not well defined, and thus the resulting network lacks the necessary correspondence or interpretability that $X_k \not\perp\!\!\!\perp X_j | X_{V \setminus \{k,j\}} \iff \theta_{kj} \neq 0$ and $\theta_{jk} \neq 0$.

Jia et al. [74] addressed this issue by introducing a random effect model-based transformation for RNA-seq data, which transforms count data to continuous data, then applying a semiparametric transformation [103] to further transform the data to be Gaussian, and finally employing the ψ-learning method to learn the desired GRN. Their method can be justified to be consistent for the recovery of the underlying GRN.

Conceptually, the method of Jia et al. [74] is closely related with the latent copula method [38, 65], where Gaussian latent variables are simulated in place of Poisson variables in recovering GRNs. However, as pointed out by Fan et al. [44], the conditional independence between the latent variables does not imply the conditional independence between the observed discrete variables.

5.2   Data-Continuized Transformation

Let $X_i = (X_{i1}, X_{i2}, \ldots, X_{in})^T$ denote the expression level of gene i measured in RNA-seq on n subjects. Consider a hierarchical random effect model:

$$\begin{aligned} X_{ij} &\sim \text{Poisson}(\theta_{ij}), \quad \theta_{ij} \sim \text{Gamma}(\alpha_i, \beta_i), \\ \alpha_i &\sim \text{Gamma}(a_1, b_1), \quad \beta_i \sim \text{Gamma}(a_2, b_2), \end{aligned} \qquad (5.3)$$

for $i = 1, 2, \ldots, p$ and $j = 1, 2, \ldots, n$, where a_1, b_1, a_2 and b_2 are prior hyperparameters. In this model, the gene-specific random effect is modeled by a Gamma distribution and, by integrating out θ_{ij}, X_{ij} follows a negative binomial distribution $NB(r, q)$ with $r = \beta_i$ and $q = \alpha_i/(1 + \alpha_i)$. This implies that the model (5.3) has the necessary flexibility for accommodating potential overdispersion of the RNA-seq data.

Let $x_i = \{x_{ij} : j = 1, 2, \ldots, n\}$ denote a realization of X_i. By assuming that the parameters $\{\alpha_i, \beta_i : i = 1, 2, \ldots, n\}$ are *a priori* independent, it is easy to derive

$$
\begin{aligned}
\pi(\alpha_i | \theta_{ij}, \beta_i, x_i) &\propto \frac{\alpha_i^{a_1 - 1}}{\Gamma^n(\alpha_i)} e^{\alpha_i \left(-b_1 + n \log \beta_i + \sum_{j=1}^{n} \log \theta_{ij} \right)}, \\
\pi(\beta_i | \alpha_i, \theta_{ij}, x_i) &\propto \beta_i^{n\alpha_i + a_2 - 1} e^{-\beta_i (\sum_{j=1}^{n} \theta_{ij} + b_2)}, \\
\pi(\theta_{ij} | \alpha_i, \beta_i, x_i) &\propto \theta_{ij}^{y_{ij} + \alpha_i - 1} e^{-\theta_{ij}(1 + \beta_i)},
\end{aligned}
\tag{5.4}
$$

that is, $\beta_i | \alpha_i, \theta_{ij}, x_i \sim Gamma(n\alpha_i + a_2, \sum_{j=1}^{n} \theta_{ij} + b_2)$ and $\theta_{ij} | \alpha_i, \beta_i, x_i \sim Gamma(y_{ij} + \alpha_i, \beta_i + 1)$.

For the choices of the prior hyperparameters a_1, b_1, a_2, b_2, Jia et al. [74] established the following lemma:

Lemma 5.1 *(Lemma 2.1; [74]) If a_1 and a_2 take small positive values, then for any $i \in \{1, 2, \ldots, p\}$ and $j \in \{1, 2, \ldots, n\}$,*

$$
\left| E[\theta_{ij} | x_i] - x_{ij} \right| \to 0, \quad \text{as } b_1 \to \infty \text{ and } b_2 \to \infty,
$$

where $E[\theta_{ij} | x_i]$ denotes the posterior mean of θ_{ij}.

By Lemma 5.1, Jia et al. [74] suggested to simulate from the posterior (equation 5.4) using an adaptive MCMC algorithm, where the values of b_1 and b_2 increase with iterations. In particular, they suggested to set

$$
b_1^{(t)} = b_1^{(t-1)} + \frac{c}{t^\zeta}, \quad b_2^{(t)} = b_2^{(t-1)} + \frac{c}{t^\zeta}, \quad t = 1, 2, \ldots,
\tag{5.5}
$$

where $b_i^{(t)}$, $i = 1, 2$, denotes the value of b_i at iteration t, $b_1^{(0)}$ and $b_2^{(0)}$ are large constants, $c > 0$ is a small constant, and $0 < \zeta \leq 1$. Suppose that the Metropolis-within-Gibbs sampler [121] is used in simulating from the posterior distribution (equation 5.4), where α_i is updated by the Metropolis-Hastings algorithm [62, 119] with a random walk proposal and β_i and θ_{ij} are updated by the Gibbs sampler [56]. Then the following lemma holds:

Lemma 5.2 *(Lemma 2.2; [74]) If the Metropolis-within-Gibbs sampler is used for simulating from the posterior distribution (equation 5.4) as described above, and the prior hyperparameters are gradually updated as in equation (5.5), then for any i and j,*

$$
\hat{\theta}_{ij}^{(T)} - x_{ij} \xrightarrow{p} 0, \quad \text{as } T \to \infty,
$$

where $\hat{\theta}_{ij}^{(T)} = \sum_{t=1}^{T} \theta_{ij}^{(t)} / T$ and $\theta_{ij}^{(t)}$ denotes the posterior sample of θ_{ij} simulated at iteration t.

Let $X_1, X_2, \ldots, X_p$ be generic variables for expression levels of p genes, and let $\hat{\theta}_1, \hat{\theta}_2, \ldots, \hat{\theta}_p$ denote the corresponding continued variable. Let $\mathbb{F}_{x_1, \ldots, x_p}(t)$ denote the empirical CDF of the expression levels of p genes, which can be estimated based on the samples $\{x_{ij} : i = 1, 2, \ldots, p, j = 1, 2, \ldots, n\}$; and let $\mathbb{F}_{\hat{\theta}_1^{(T)}, \ldots, \hat{\theta}_p^{(T)}}(t)$ denote the empirical CDF of the continued expression levels of p genes, which can be estimated based on $\{\hat{\theta}_{ij} : i = 1, 2, \ldots, p, j = 1, 2, \ldots, n\}$. Lemma 5.2 implies

$$\sup_{t \in \mathbb{R}^p} \|\mathbb{F}_{\hat{\theta}_1^{(T)}, \ldots, \hat{\theta}_p^{(T)}}(t) - \mathbb{F}_{x_1, \ldots, x_p}(t)\| \xrightarrow{P} 0, \quad \text{as } T \to \infty.$$

On the other hand, under some regularity and sparsity conditions (for the underlying graphical model formed by $X_1, X_2, \ldots, X_p$),

$$\sup_{t \in \mathbb{R}^p} \|\mathbb{F}_{x_1, \ldots, x_p}(t) - F_{X_1, \ldots, X_p}(t)\| \xrightarrow{a.s.} 0, \quad \text{as } n \to \infty,$$

where $F_{X_1, \ldots, X_p}(t)$ denotes the CDF of X_i's. Simply combining the two limits leads to

$$\sup_{t \in \mathbb{R}^p} \|\mathbb{F}_{\hat{\theta}_1^{(T)}, \ldots, \hat{\theta}_p^{(T)}}(t) - F_{X_1, \ldots, X_p}(t)\| \xrightarrow{P} 0, \quad \text{as } T \to \infty \text{ and } n \to \infty,$$

which implies that the continuized data can replace the observed data in studying the dependency relationships among the variables $X_1, \ldots, X_p$ under appropriate conditions.

5.3   Data-Gaussianized Transformation

Let $X = (X_1, \ldots, X_p)^T$ denote a p-dimensional continuous random vector.

Definition 5.1 *(Nonparanormal distribution, [103]) It is said that X follows a nonparanormal distribution, i.e., $X \sim NPN(\mu, \Sigma, f)$, if there exist functions $\{f_j\}_{j=1}^p$ such that $Z = (Z_1, Z_2, \ldots, Z_p)^T = f(X) \sim N(\mu, \Sigma)$, where $f(X) = (f_1(X_1), \ldots, f_p(X_p))^T$. In particular, if f_j's are monotone and differentiable, the joint probability density function of X is given by*

$$P_X(x) = \frac{1}{(2\pi)^{p/2} |\Sigma|^{1/2}} \exp\left\{ -\frac{1}{2}(f(x) - \mu)^T \Sigma^{-1} (f(x) - \mu) \right\} \prod_{j=1}^{p} |f_j'(x_j)|. \quad (5.6)$$

Based on equation (5.6), Liu et al. [103] showed $X_i \perp X_j | X_{V \setminus \{i,j\}} \iff Z_i \perp Z_j | Z_{V \setminus \{i,j\}}$, provided that $X \sim NPN(\mu, \Sigma, f)$ and each f_j is monotone and differentiable. That is, the nonparanormal transformation preserves the conditional independence relationships among the variables $X_1, X_2, \ldots, X_p$. Liu et al. [103] further showed that the nonparanormal transformation can be made by setting

$$f_j(x) = \mu_j + \sigma_j \Phi^{-1}(F_j(x)), \quad j = 1, 2, \ldots, p, \quad (5.7)$$

where μ_j, σ_j^2 and $F_j(x)$ denote, respectively, the mean, variance, and CDF of X_j. To robustify the transformation, $F_j(x)$ can be replaced by a truncated or Winsorized estimator of the marginal empirical CDF of X_j, especially under the high-dimensional scenario.

5.4 Learning Gene Regulatory Networks with RNA-seq Data

With the data-continuized and data-Gaussianized transformations and the ψ-learning method, Jia et al. [74] proposed the following procedure for learning GRN with NGS data:

Algorithm 5.1 *(ψ-Learning with RNA-seq data; [74])*

a. Apply the data-continuized transformation to continuized RNA-seq data.

b. Apply the nonparanormal transformation to continuized RNA-seq data.

c. Apply the ψ-learning method to Gaussianized RNA-seq data.

The consistency of the method in GRN recovery follows from Lemma 5.2, invariance of the conditional independence relations with respect to nonparanormal transformations, and consistency of the ψ-learning method.

5.5 Application

This example is on the mRNA sequencing data collected from acute myeloid leukemia (AML) patients,[1] which consisted of 179 subjects and 19,990 genes. The dataset has been analyzed by Jia et al. [74], and their results are briefly reviewed as follows.

Jia et al. [74] preprocessed the data by first filtering out the low expression genes by excluding those with one or more zero counts and then selecting 500 genes with the highest inter-sample variation. From the GRN learned by the method, they identified a few hub genes such as MKI67 and KLF6, which are known to be associated with acute myeloid leukemia [69, 120].

For comparison Jia et al. [74] applied the graphical Lasso [49], nodewise regression[2] [118], and local Poisson graphical modeling[3] (LPGM) [3] methods to the preprocessed data. They found that the network learned by Algorithm 5.1 approximately follows a power-law distribution (see Appendix A.3), while those learned by other methods do not. This partially justifies the validity of Algorithm 5.1. It is known that many biological networks follow the power-law distribution, see e.g. [7] for discussions on this issue.

5.6 Problems

1. Provide the detail for the derivation of equation (5.2).

2. Provide the detail for the derivation of equation (5.4).

3. Prove Lemma 5.1.

4. Prove Lemma 5.2.

---

[1]The dataset were available at *The Cancer Genome Atlas* (TCGA) data portal http:// cancergenome.nih.gov/.

[2]The graphical Lasso and nodewise regression methods were implemented with the R package *huge* [200] under the default settings except that the regularization parameters were selected by the stability approach.

[3]It was implemented with the package *XMRF* [178] and run under its default setting.

5. Prove that the nonparanormal transformation (equation 5.7) preserves the conditional independence relationships among the continuous random variables $X_1, X_2, \ldots, X_p$.

6. Perform a simulation study for Algorithm 5.1.

Chapter 6

Mixed Graphical Models

This chapter describes the p-learning method [183] for learning pairwise mixed graphical models. The p-learning method is developed based on the theory of Bayesian network, more precisely, the property of Markov blanket. As discussed in the chapter, Markov blanket is a fundamentally important concept for statistical analysis of high-dimensional data. It can lead to a significant reduction of the dimension of the data, enabling many statistical tests, e.g., t-test and Wald-test, that are originally developed for low-dimensional problems to be used for high-dimensional statistical inference.

6.1 Introduction

Many datasets collected in modern data science contain both continuous and discrete types of variables. For example, a genomic dataset might contain gene expression data and mutation data; although the former is (for microarray data) or can be transformed to be continuous (see Chapter 5 for the transformation of RNA-seq data), the latter is categorical. Another example is social survey data, where the variables are often mixed, including both continuous variables (e.g., household income and age) and discrete variables (e.g., gender and occupation).

Regarding graphical modeling for mixed data, quite a few works have been published in the literature. A nonexhaustive list includes [22, 23, 44, 47, 88, 160, 183, 186]. Among them, Chen et al. [22], Cheng et al. [23], Fellinghauer et al. [47], Lee and Hastie [88], Sun and Liang [160], and Yang et al. [186] studied the joint and conditional distributions of the mixed graphical model and estimated the parameters of the model using nodewise regression, maximum pseudo-likelihood or graphical random forest methods. Fan et al. [44] proposed a latent variable method, which imputes latent Gaussian variables for the discrete variables and then estimates the Gaussian graphical model using a regularization method. Xu et al. [183] proposed a p-learning method based on the theory of Bayesian networks and the theory of sure independence screening for generalized linear models (GLMs) [46]. The p-learning method was developed in the same vine as the ψ-learning method under the paradigm of conditional independence test.

In what follows, we first give a brief review of mixed graphical models, mainly focusing on their joint and conditional distributions; then give a brief review of Bayesian networks, mainly focusing on their network structure; and finally describe

DOI: 10.1201/9780429061189-6

the p-learning method. The straightforward goal of the p-learning method is to learn the moral graph for a Bayesian network, while the moral graph is an undirected Markov network and is exactly the goal pursued by other methods of mixed graphical modeling.

6.2 Mixed Graphical Models

Consider a dataset that consists of both continuous and discrete random variables. Let $\{Y_1, Y_2, \ldots, Y_{p_c}\}$ denote the set of continuous random variables, let $\{Z_1, Z_2, \ldots, Z_{p_d}\}$ denote the set of discrete random variables, and let $p = p_c + p_d$. For simplicity, let's restrict our considerations to pairwise graphical models, i.e., all the variables in the model are only allowed to interact pairwise. The joint and conditional distributions for such models have been studied in quite a few works, see e.g., [22, 88]. For example, for a mixed graphical model of Gaussian and multinomial random variables [88], gave the following joint distribution:

$$
\begin{aligned}
p(y,z;\Theta) \propto \exp\Big\{ &-\frac{1}{2}\sum_{s=1}^{p_c}\sum_{t=1}^{p_c}\theta_{st}y_sy_t + \sum_{s=1}^{p_c}\vartheta_s y_s + \sum_{s=1}^{p_c}\sum_{j=1}^{p_d}\rho_{sj}(z_j)y_s \\
&+ \sum_{j=1}^{p_d}\sum_{r=1}^{p_d}\psi_{rj}(z_r,z_j)\Big\},
\end{aligned}
\tag{6.1}
$$

where $\Theta = [\{\theta_{st}\}, \{\vartheta_s\}, \{\rho_{sj}\}, \{\psi_{rj}\}]$ denotes the parameter set of the joint distribution. Suppose that Y_r has K_r states for $r = 1, 2, \ldots, p_d$. Then the elements of Θ can be interpreted as follows: θ_{st} is the continuous–continuous edge potential, ϑ_s is the continuous node potential, $\rho_{sj}(z_j)$ is the continuous-discrete edge potential taking values from a set of K_j values, and $\psi_{rj}(z_r,z_j)$ is the discrete–discrete edge potential taking values from a set of $K_r \times K_j$ values.

Following Lee and Hastie [88], one can further derive that the conditional distribution of Y_r given all other variables is multinomial with the probability distribution given by a multiclass logistic regression:

$$
p(z_r = k|z_{\backslash r}, y; \Theta) = \frac{\exp(\omega_{0k} + \sum_j \omega_{kj}x_j)}{\sum_{l=1}^{K_r}\exp(\omega_{0l} + \sum_j \omega_{lj}x_j)},
\tag{6.2}
$$

where $z_{\backslash r}$ denotes the set of discrete variables except for z_r, ω_{lj} are regression coefficients, x_j is defined as a dummy variable for a discrete variable, e.g., $x_j = 1_{z_u=k}$, and $x_j = y_u$ for a continuous variable. The conditional distribution of Y_s given other variables is Gaussian with the probability density function given by a Gaussian linear regression:

$$
p(y_s|y_{\backslash s}, z; \Theta) = \frac{1}{\sqrt{2\pi}\sigma_s}\exp\left(-\frac{1}{2\sigma_s^2}(y_s - \omega_0 - \sum_j \omega_j x_j)^2\right),
\tag{6.3}
$$

where x_j is as defined in equation (6.2).

In a slightly more general form, Chen et al. [22] specified each conditional distribution as a member from the exponential family:

$$p(x_s | x_{\setminus s}) = \exp \left\{ f_s(x_s) + \sum_{t \neq s} \theta_{ts} x_t x_s - D_s(\eta_s) \right\}, \quad s = 1, 2, \ldots, p, \qquad (6.4)$$

where x_s represents a generic variable, continuous or discrete, in the mixed graphical model; $f(x_s) = \alpha_{1s} x_s + \alpha_{2s} x_s^2 / 2 + \sum_{k=3}^{K} \alpha_{ks} B_{ks}(x_s)$ is the node potential defined with some known functions $B_{ks}(\cdot)$'s; and $\eta_s = \alpha_s + \sum_{t:t \neq s} \theta_{ts} x_t$. Further, if $\theta_{st} = \theta_{ts}$, the conditional distributions are compatible and the joint distribution is given by

$$p(x_1, x_2, \ldots, x_p) = \exp \left\{ \sum_{s=1}^{p} f_s(x_s) + \frac{1}{2} \sum_{s=1}^{p} \sum_{t \neq s} \theta_{ts} x_t x_s - A_{\theta, \alpha} \right\}, \qquad (6.5)$$

where $A_{\theta, \alpha}$ is the log-normalizing constant as a function of θ_{ts}'s and α_{ks}'s. Chen et al. [22] considered the mixed graphical models that $(x_1, x_2, \ldots, x_p)$ are mixed with Gaussian, Bernoulli, Poisson, and exponential random variables or a subset of them. To ensure the joint distribution to be proper, i.e., $A_{\theta, \alpha}$ is finite, θ_{ts}'s need to be restricted to different ranges for different pairs of random variables. For example, θ_{ts} needs to be non-positive for the pairs of Poisson-Poisson, Poisson-exponential, and exponential-exponential. Refer to [22] for the detail.

Given the above conditional distributions, a natural method for determining the structure of the mixed graphical model is nodewise regression. Chen et al. [22] established the consistency of the method for the case that each regression is penalized by a Lasso penalty [160, 164], extended the theory to the case of amenable penalties [105, 106], which has included the Lasso penalty [164], SCAD penalty [43], and MCP penalty [195] as special cases.

Other than the above pairwise graphical models, Cheng et al. [23] studied the models with third-order interaction terms, and Yang et al. [186] studied the models of block directed Markov random fields where the variables are mixed and the edges can be directed or undirected.

6.3 Bayesian Networks

Consider a set of random variables $X = \{X_1, X_2, \ldots, X_p\}$, which can consist of both continuous and discrete random variables. A Bayesian network for X, as illustrated by Figure 6.1a, is a directed acyclic graph (DAG) represented by $G = (V, E)$, where $V := \{1, 2, \ldots, p\}$ is a set of p nodes with each node corresponding to a variable in X, and $E = (e_{ij})$ is the adjacency matrix that specifies the edges and edge directions of the DAG. For example, if $e_{ij} = 1$ then there is an directed edge from node X_i to node X_j, which can also be represented as "$X_i \rightarrow X_j$". The joint distribution of X can be specified through the Bayesian network as follows:

$$P(X) = \prod_{i=1}^{p} q(X_i | Pa(X_i)), \qquad (6.6)$$

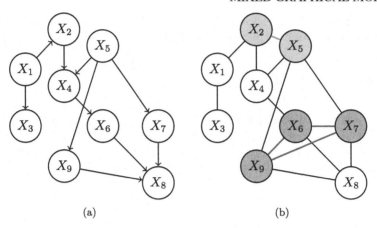

Figure 6.1 *A Bayesian network (a) and its moral graph (b) obtained by adding the edges* $X_2 - X_5$, $X_6 - X_7$, $X_6 - X_9$ *and* $X_7 - X_9$.

where $Pa(X_i)$ denotes the set of parent variables of X_i, and $q(X_i|Pa(X_i))$ is the conditional distribution of X_i given the parent variables. The decomposition (equation 6.6) can be viewed as a sparse representation of the general chain decomposition form

$$P(X) = \prod_{i=1}^{p} q(X_i|X_{i-1}, \ldots, X_1),$$

based on the generative structure of the Bayesian network.

6.3.1   Basic Terms

To have a better explanation for the concept of Bayesian networks, we define some terms that are frequently used in this chapter. Refer to [72, 142] for a full development of the theory for Bayesian networks.

Definition 6.1 *(Collider) The two edges* $X_i \rightarrow X_k \leftarrow X_k$ *form a collider (also known as V-structure) if there is no edge connecting* X_i *and* X_j, *and* X_k *is called a collision node.*

With colliders, Bayesian networks can represent a type of relationship that undirected Markov networks cannot, i.e., X_i and X_j are marginally independent, while they are conditionally dependent given X_k. This property has been considered as an advantage of Bayesian networks over undirected Markov networks in variable dependency representation [129].

Definition 6.2 (Descendant) *A variable* X_j *is called a descendant of another variable* X_i *if there is a path connecting from* X_i *to* X_j.

Definition 6.3 (Local Markov property) *In Bayesian networks, each variable* X_i *is conditionally independent of its non-descendant variables given its parent variables.*

The local Markov property forms the basis of statistical inference with Bayesian networks. As exampled by the collider, this property implies that a variable is not completely independent of its descendant variables. The dependence caused by the descendants can be easily understood from Bayesian theorem. For example, let's consider a Bayesian network of three nodes, $X_1 \rightarrow X_3 \leftarrow X_2$, the associated Bayesian theorem is given by

$$\pi(X_1, X_2 | X_3) = \frac{\pi(X_3 | X_1, X_2) \pi(X_1) \pi(X_2)}{\pi(X_3)} \neq \pi(X_1) \pi(X_2),$$

which provides a mathematical formulation for how the conditional distribution of the parent variables changes with the information of the child variables.

Definition 6.4 (Markov blanket) *The Markov blanket of a variable X_i is the set of variables that consist of the parent variables of X_i, the child variables of X_i, and the spouse variables that share child variables with X_i.*

The Markov blanket of X_i is the minimal subset of $\{X_1, X_2, \ldots, X_p\}$ conditioned on which X_i is independent of all other variables. This property is fundamentally important for high-dimensional statistical inference, with which, as shown in Chapter 9, the high-dimensional statistical inference problem can be reduced to a series of low-dimensional statistical inference problems.

Definition 6.5 (Moral graph) *The moral graph is an undirected Markov network representation of the Bayesian network, which, as illustrated by Figure 6.1b, is constructed by (i) connecting the non-adjacent nodes in each collider by an undirected edge, and (ii) removing the directions of other edges.*

The transformation from a Bayesian network to its moral graph is called moralization. In the moral graph, all dependencies are explicitly represented, and the neighboring set of each variable forms its own Markov blanket. Another important term for Bayesian networks is *faithfulness*, for which the definition given in Chapter 2 still holds. The faithfulness forms the theoretical basis for establishing consistency of the *p*-learning method by Xu et al. [183].

In the proof of Theorem 6.1, a term "separator" is involved. This term has been defined for undirected Markov networks in Chapter 2. For Bayesian networks, its definition is slightly different due to the existence of colliders.

Definition 6.6 (Separator) *For a Bayesian network, a subset U is said to be a separator of the subsets I and J if for any node X_i in I and any node X_j in J, every path between X_i and X_j satisfies the conditions: (i) the path contains the chain connections $X_a \rightarrow X_b \rightarrow X_c$ or the fork connections $X_a \leftarrow X_b \rightarrow X_c$ such that X_b is included in U, or (ii) the path contains the collider connections $X_a \rightarrow X_b \leftarrow X_c$ such that neither the collision node X_b nor any of its descendants are included in U.*

It is worth pointing out that Bayesian networks are different from causal Bayesian networks in structural interpretation, although both are DAGs. For causal Bayesian networks, each edge represents a direct cause-effect relation between the parent variable and the child variable, while this might not hold for Bayesian networks. To learn a causal Bayesian network, the data from experimental interventions are generally needed and it is difficult, if not impossible, by using observational data alone. See [31, 36, 42] for more discussions on this issue and some exceptional cases.

6.3.2 Bayesian Network Learning

The existing methods for learning the structure of a Bayesian network can be roughly grouped into three categories, namely, constraint-based methods, score-based methods, and hybrid methods.

The constraint-based methods are to learn Bayesian networks by conducting a series of conditional independence tests. There are a variety of methods developed in this category. To name a few, they include inductive causation [176], PC [153], grow-shrink [110], incremental association Markov blanket [168, 189], and semi-interleaved HITON-PC [2], among others. The constraint-based methods generally work in three steps:

a. Learning a moral graph through a series of conditional independence tests.
b. Identifying the colliders in the moral graph, see e.g., [111, 130].
c. Identifying the derived directions for non-convergent edges according to logic rules, see e.g., [80, 177].

A general difficulty with the constraint-based methods is that they are not well-scaled with respect to the dimension [2]: They often involve some conditional tests with a large conditioning set, making the methods slow and unreliable especially when p is much greater than n. The p-learning method [183] belongs to the category of constraint-based methods, and it overcomes the scalability issue through a sure screening step that significantly reduces the size of the conditioning set of the tests.

The score-based methods are to find a Bayesian network that minimizes an appropriately chosen score function, which evaluates the fitness of each feasible network to the data. Examples of such score functions include Bayesian score [30, 100] and minimum description length [84], among others. The score-based methods can be shown to be consistent under appropriate conditions, see e.g., [26, 122]. Unfortunately, minimizing the score function is known to be NP-hard [25], and the search process often stops at a sub-optimal solution as the surface of the score function can be rugged. This is particularly true when the dimension p is high.

The hybrid methods are proposed as a combination of the constraint-based and score-based methods, which are first to restrict the parent set of each variable to be of a small size and then search for the Bayesian network that minimizes a selected score function under the restriction, see e.g., sparse candidate [52] and max-min hill-climbing [167].

6.4 p-Learning Method for Mixed Graphical Models

As a constraint-based method, the p-learning method [183] learns the moral graph through a series of conditional independence tests under the assumption of faithfulness. To improve the power and scalability (with respect to the dimension p) of the conditional independence tests, the *p*-learning method reduces the sizes of their conditioning sets through a sure independence screening procedure based on the property of Markov blanket.

Consider a pair of random variables (X_i, X_j). To determine the edge status between the two variables/nodes, one needs to conduct the conditional independence test $X_i \perp\!\!\!\perp X_j | X_{V \setminus \{i,j\}}$. Let S_i and S_j denote the Markov blankets of X_i and X_j, respectively. By the total conditioning property of Bayesian networks [130] and the property of Markov blanket,

$$X_i \not\!\perp\!\!\!\perp X_j | X_{S_i \setminus \{i,j\}} \iff j \in S_i \iff X_i \not\!\perp\!\!\!\perp X_j | X_{V \setminus \{i,j\}}, \tag{6.7}$$

which implies that the test $X_i \perp\!\!\!\perp X_j | X_{V \setminus \{i,j\}}$ can be reduced to $X_i \perp\!\!\!\perp X_j | X_{S_i \setminus \{i,j\}}$ if the Markov blanket S_i is known. By the symmetry of Markov blanket, i.e., $i \in S_j \iff j \in S_i$, the test $X_i \perp\!\!\!\perp X_j | X_{V \setminus \{i,j\}}$ can also be reduced to $X_i \perp\!\!\!\perp X_j | X_{S_j \setminus \{i,j\}}$ if S_j is known. However, for a given dataset, the Markov blanket of each variable is generally unknown. To address this issue, Xu et al. [183] proposed to find a super Markov blanket $\tilde{S}_i (\supseteq S_i)$, instead of the exact Markov blanket, for each variable X_i; and they showed that they are equivalent for the conditional independence test in the sense of equation (6.8). That is, the super and exact Markov blankets are equivalent in recovering the structure of the moral graph.

Let ϕ_{ij} be the indicator of the test $X_i \perp\!\!\!\perp X_j | X_{S_i \setminus \{i,j\}}$, i.e., $\phi_{ij} = 1$ if the conditional independence holds and 0 otherwise. Let $\tilde{\phi}_{ij}$ be the indicator of the test $X_i \perp\!\!\!\perp X_j | X_{\tilde{S}_i \setminus \{i,j\}}$. Theorem 6.1 shows that ϕ_{ij} and $\tilde{\phi}_{ij}$ are equivalent in recovering the structure of the moral graph under the faithfulness assumption.

Theorem 6.1 (*Theorem 1; [183]*) *Assume the faithfulness holds for a Bayesian network. Then* ϕ_{ij} *and* $\tilde{\phi}_{ij}$ *are equivalent in recovering the structure of the moral graph in the sense that*

$$\phi_{ij} = 1 \iff \tilde{\phi}_{ij} = 1. \tag{6.8}$$

Proof 2 *If* $\phi_{ij} = 1$, *then* $S_i \setminus \{i,j\}$ *forms a separator of* X_i *and* X_j. *Since* $S_i \subset \tilde{S}_i$, $\tilde{S}_i \setminus \{i,j\}$ *is also a separator of* X_i *and* X_j. *By faithfulness, we have* $\tilde{\phi}_{ij} = 1$.

On other hand, if $\tilde{\phi}_{ij} = 1$, *then* X_i *and* X_j *are conditionally independent and* $\tilde{S}_i \setminus \{i,j\}$ *forms a separator of* X_i *and* X_j. *Since* $\tilde{S}_i \subset V$, $V \setminus \{i,j\}$ *is also a separator of* X_i *and* X_j *and the conditional independence* $X_i \perp\!\!\!\perp X_j | X_{V \setminus \{i,j\}}$ *holds. By the property (equation 6.7),* $j \notin S_i$ *in this case. Therefore,* $\phi_{ij} = 1$ *holds.*

By the symmetry of Markov blanket, Theorem 6.1 also holds if $\tilde{S}_i$ is replaced by $\tilde{S}_j$. Considering the power of the test, the smaller of $\tilde{S}_i$ or $\tilde{S}_j$ is generally preferred. This formulation of conditional independence tests simplifies the formulation (equation 1.7) based on the property of Bayesian networks.

Based on Theorem 6.1, Xu et al. [183] proposed the following algorithm for simultaneously learning a super Markov blanket for each variable.

Algorithm 6.1 (*p-learning method; [183]*)

a. (*Parent and child variables screening*) *Find a superset of parent and child variables for each variable* X_i:

    i. *For each ordered pair (X_i, X_j), perform the marginal independence test $X_i \perp\!\!\!\perp X_j$ to get a p-value.*

   ii. *Perform a multiple hypothesis test (at a significance level of α_1) on the p-values to identify the pairs of variables that are marginally dependent. Denote the set of variables that are marginally dependent on X_i by A_i for $i = 1, \ldots, p$. If $|A_i| > n/(c_{n1} \log(n))$ for a pre-specified constant c_{n1}, where $|A_i|$ denotes the cardinality of A_i, reduce it to $n/(c_{n1} \log(n))$ by removing the variables having larger p-values in the marginal independence tests.*

 b. *(Spouse variable amendment) For each variable X_i, find the spouse variables that are not included in A_i, i.e., finding $B_i = \{j : j \notin A_i, \exists\ k \in A_i \cap A_j\}$. If $|B_i| > n/(c_{n2} \log(n))$ for a pre-specified constant c_{n2}, reduce it to $n/(c_{n2} \log(n))$ by removing the variables having larger p-values in the spouse test $X_i \perp\!\!\!\perp X_j | X_{A_i \cap A_j \setminus \{i,j\}}$.*

 c. *(Moral graph recovery) Determine the structure of the moral graph through conditional independence tests:*

    i. *For each ordered pair (X_i, X_j), perform the conditional independence test $X_i \perp\!\!\!\perp X_j | X_{\tilde{S}_{ij} \setminus \{i,j\}}$, where $\tilde{S}_{ij} = A_i \cup B_i$ if $|A_i \cup B_i \setminus \{i,j\}| \leq |A_j \cup B_j \setminus \{i,j\}|$ and $\tilde{S}_{ij} = A_j \cup B_j$ otherwise.*

   ii. *Perform a multiple hypothesis test at a significance level of α_2 to identify the pairs of variables that are conditionally dependent, and set the adjacency matrix $\hat{E}_{mb}$ accordingly, where $\hat{E}_{mb}$ denotes the adjacency matrix of the moral graph.*

For each variable X_i, step (a) finds a superset of parent and child variables and step (b) finds the spouse variables, therefore,

$$S_i \subset \tilde{S}_i := A_i \cup B_i,$$

which, by Theorem 6.1, implies that the algorithm is valid for moral graph recovery. This algorithm is called p-learning method as the screening step is p-value based, compared to the ψ-learning method described in Chapter 2.

    Throughout this chapter, it is assumed that the conditional distribution of each node belongs to the exponential family, for which the probability mass/density function can be expressed in the form (equation 6.4). Therefore, by the sure screening property of GLMs [46], the marginal independence test in step (a) can be performed by testing whether or not the coefficient of X_j in the following GLM is equal to zero:

$$X_i \sim 1 + X_j. \tag{6.9}$$

Similarly, the conditional independence test in step (c) can be performed by testing whether or not the coefficient of X_j in the following GLM is equal to zero:

$$X_i \sim 1 + X_j + \sum_{k \in \tilde{S}_{ij} \setminus \{i,j\}} X_k. \tag{6.10}$$

The multiple hypothesis tests in steps (a) and (c) can be done using the empirical Bayes method developed in Liang and Zhang [95], which allows for general dependence between test statistics. Also, a slightly larger value of α_1, e.g., 0.05, 0.1 or 0.2, is generally preferred. The choice of α_2 depends on the sparsity level of the network one preferred. Throughout this chapter, $\alpha_1 = 0.1$ and $\alpha_2 = 0.05$ are set to the default unless stated otherwise. In practice, one might always conduct the spouse tests to reduce the sizes of B_i's even if they are small, as the sizes of $\tilde{S}_i$'s can adversely affect the power of the moral graph screening test.

6.5 Simulation Studies

For simulating a dataset from a mixed graphical model, two methods can be used. The first one is based on the conditional distribution of each variable; that is, one can employ a Gibbs sampler [56] to iteratively draw samples according to the conditional distribution (equation 6.4). This method has been used in, e.g., [22].

The other method is based on a temporal order of the nodes (see e.g., [51]). For any DAG, there exists a temporal order of the nodes such that for any two nodes X_i and X_j, if there is an edge $X_i \rightarrow X_j$, then X_i must be preceding to X_j in the temporal order. Based on the temporal order, the dataset can be simulated in the following procedure [78]: (i) specify a temporal order of variables, and randomly mark each variable as continuous or binary with an equal probability and then (ii) generate the data according to a pre-specified edge set in a sequential manner.

In our simulation studies, the directed edge set was specified by

$$e_{ij} = \begin{cases} 1, & \text{if } j - i = 1 \text{ or } 2, \\ 0, & \text{otherwise,} \end{cases} \tag{6.11}$$

for any $j > i$, and the data were generated in the following procedure. We first generated $Z_1 \sim N(0,1)$, and set

$$X_1 = \begin{cases} Z_1, & \text{if } X_1 \text{ was marked as continuous,} \\ \text{Bernoulli}(1/(1+e^{-Z_1})), & \text{otherwise,} \end{cases}$$

where $\text{Bernoulli}(1/(1+e^{-Z_1}))$ denotes a Bernoulli random variable with the success probability $1/(1+e^{-Z_1})$; then we generated $X_2, X_3, \ldots, X_p$ by setting

$$Z_j = \frac{1}{2} \sum_{i=1}^{j} e_{ij} X_i,$$

$$X_j = \begin{cases} Z_j + \varepsilon_j, & \text{if } X_j \text{ was marked as continuous,} \\ \text{Bernoulli}(1/(1+e^{-Z_j})), & \text{if } X_j \text{ was marked as binary,} \end{cases}$$

where $\varepsilon_j \sim N(0,1)$. We tried different settings of n and (p_c, p_d). For each setting, we simulated ten datasets independently. The results of the p-learning method are reported in Table 6.1, which indicates that the method performs robustly to the dimensions (p_c, p_d).

For comparison, we have applied some constraint-based methods such as PC stable [29], grow-shrink (GS) [110], incremental association Markov blanket (IAMB)

Table 6.1 *Average areas under the ROC curves (over ten independent datasets) produced by the p-learning method, where the number in the parenthesis represents the standard deviation of the average area.*

(p_c, p_d) n	(50,50)	(100,100)	(150,150)
200	0.8954 (0.0085)	0.8843 (0.0039)	0.8835 (0.0025)
500	0.9433 (0.0017)	0.9458 (0.0024)	0.9430 (0.0018)

[168], and semi-interleaved HITON-PC (si.hiton.pc) [2] to the simulated data. These methods have been implemented in the R package *bnlearn* [143], but they are too slow to be applied for the conditional independence tests with a high significance level. In consequence, the AUC values under their precision-recall or receiver-operating-characteristic (ROC) curves cannot be reliably evaluated and thus not reported here. Xu et al. [183] calculated the AUC values of the ROC curves by making a simple linear line extrapolation from a point of low false positive rate to the corner (1,1). Refer to [183] for the detail.

The mixed graphical model can also be learned with nodewise regression method, see e.g., [22, 160], but this is not the focus of this book.

6.6   Application

This example is on a dataset downloaded from The Cancer Genome Atlas (TCGA),[1] which consists of microarray mRNA gene expression data, mutation data and DNA methylation data. The gene expression data have been normalized as Gaussian, and the mutation and methylation data have been coded as binary. This dataset has been analyzed by Xu et al. [183] and their results are briefly reviewed as follows.

Xu et al. [183] restricted the analysis to the genes included in the BRCA pathways, which can be found at Kyoto Encyclopedia of Genes and Genomes (KEGG). By this restriction, they reduced the dataset to $n = 287$ samples, each sample consisting of 129 mRNA gene expressions, 11 mutations, and 315 DNA methylations. In addition, by biological knowledge, they exempted three types of edges: mutation-mutation, methylation-methylation, and mutation-methylation. For the *p*-learning method, this is simple: one can just work on the pairs of variables for which the edge is not exempted in steps (a) and (c) of Algorithm 6.1.

From the moral graph learned by the *p*-learning method, Xu et al. [183] identified some hub genes, mutations and methylations. They found that their results are consistent with the literature. For example, they identified TP53 as a hub mutation, which is known as the most common genetic alternation for breast cancer [13, 148]. Refer to [183] for a detailed report on their results.

---

[1]The TCGA database is available at https://tcga-data.nci.nih.gov/tcga/.

6.7 Consistency of the p-Learning Method

Under the assumption of pairwise Gaussian-multinomial graphical models, the consistency of the p-learning method can be justified based on the sure independence screening theory of GLMs [46] in a similar way to the ψ-learning method as described in Chapter 2. To make the justification, Xu et al. [183] made the following assumptions:

Assumption 6.1 *(Faithfulness) The Bayesian network is faithful with the joint Gaussian-multinomial distribution given by equation (6.1).*

Assumption 6.2 *(Dimensionality) $p_n = O(\exp(n^\delta))$, where $0 \le \delta < (1 - 2\kappa)\alpha/(\alpha + 2)$ for some constants $0 < \kappa < 1/2$ and $\alpha > 0$, p_n denotes the number of variables in the Bayesian network, and n denotes the sample size.*

Assumption 6.3 *(Sparsity) The maximum blanket size $\tilde{q}_n = \max_{1 \le i \le p_n} |S_i|$ is of order $O(n^b)$ for some constant $0 \le b < (1 - 2\kappa)\alpha/(\alpha + 2)$.*

Assumption 6.4 *(Identifiability) The regression coefficients satisfy*

$$\inf \left\{ |\beta_{ij|C}|; \beta_{ij|C} \ne 0,\ i, j = 1, 2, \ldots, p_n,\ C \subseteq \{1, 2, \ldots, p_n\} \setminus \{i, j\}, \right.$$
$$\left. |C| \le O(n/\log(n)) \right\} \ge c_0 n^{-\kappa},$$

for some constant $c_0 > 0$, where κ is as defined in Assumption 6.2, and $\beta_{ij|C}$ is the true regression coefficient of X_j in the GLMs (equations 6.9 and 6.10).

To ensure the sparsity assumption to be satisfied, Xu et al. [183] further gave some conditions for the GLMs by following the theory of Fan and Song [46]. Under these assumptions, Xu et al. [183] showed that the moral graph can be consistently identified by the p-learning method, i.e.,

$$P(\hat{E}_{m,n} = E_m) \to 1, \quad \text{as } n \to \infty, \tag{6.12}$$

where E_m denotes the edge set of the true moral graph and $\hat{E}_{m,n}$ denotes the p-learning estimator of E_m. Refer to [183] for the proof.

For other types of mixed graphical models, e.g., those with Poisson or exponential random variables in the presence, we suggest to make appropriate transformations for the data to avoid the constraints imposed on the parameter space of the graphical model [22]. For Poisson data, we suggest the data-continuized transformation, and nonparanormal transformation as discussed in Chapter 5. For exponential data, we suggest nonparanormal transformation. These transformations will facilitate the recovery of the structure of the graphical model.

6.8 Problems

1. Provide the detail for the derivations of (equations 6.2 and 6.3).

2. Prove the consistency of the p-learning method, i.e., equation (6.12).

3. Redo the simulation study in Section 6.5.

4. For the simulation study in Section 6.5, identify the colliders from the moral graph learned by the p-learning method using the algorithm of Pellet and Elisseeff [130].

5. Modify Algorithm 6.1 by replacing step (a) by nodewise Lasso regression. Redo the simulation study with different choices of regularization parameters for the nodewise Lasso regression and compare the results.

6. Extend the modified algorithm in Problem 5 to other types of mixed graphical models [186].

Chapter 7

Joint Estimation of Multiple
Graphical Models

This chapter introduces a fast hybrid Bayesian integrative analysis (FHBIA) method [77] for joint estimation of multiple Gaussian graphical models. Compared to other methods for joint estimation of multiple graphical models, FHBIA distinguishes itself from the peers by (i) Bayesian modeling of transformed ψ-scores instead of original data and (ii) integrating the data information under distinct conditions through an explicit meta-analysis step. The former makes it more efficient in computation, while the latter makes it more effective in data integration, compared to other methods. The FHBIA method provides a general framework for the joint estimation of multiple graphical models, and it has been extended to the joint estimation of multiple mixed graphical models [75].

7.1 Introduction

In Chapter 2, we considered the problem of Gaussian graphical modeling under the assumption that the data are drawn from a single Gaussian distribution. In Chapter 4, we considered the problem of Gaussian graphical modeling under the assumption that the data are drawn from a mixture Gaussian distribution, but different components of the mixture distribution share the same network structure. However, in practice, it is often the case that the data are collected from different Gaussian distributions under distinct conditions, and each distribution has a different network structure. For example, in genetic disease studies, we have often gene expression data collected from both patients and control subjects, and the patient data might be collected at different disease stages or from different tissues. For these datasets, instead of estimating a single graphical model through a data integration approach, we are interested in estimating one graphical model under each condition. This will allow us to visualize how the network structure changes along with disease progression and thus improve our understanding of the mechanism underlying the disease.

Under the paradigm of conditional independence tests, Jia and Liang [75] and Jia et al. [77] developed a fast hybrid Bayesian integrative analysis (FHBIA) method for joint estimation of multiple graphical models, where the similarity between different graphs is enhanced by a Bernoulli process prior that allocates a low probability on edge alternations between different graphs.

DOI: 10.1201/9780429061189-7

7.2   Related Works

In the literature, methods for joint estimation of multiple Gaussian graphical models have been developed from both frequentist and Bayesian perspectives. The frequentist method works by maximizing a penalized likelihood function, where the penalty function is chosen to enhance the similarity between the graphs learned under distinct conditions. For example, Danaher et al. [34] proposed to work with a fused lasso or group lasso penalty that enhances the similarity of the precision matrices of the graphical models. More precisely, they estimated the precision matrices $\{\Theta_k : k = 1, 2, \ldots, K\}$ through solving the optimization problem:

$$\max_{\{\Theta_k : k=1,2,\ldots,K\}} \left(\sum_{k=1}^{K} n_k [\log \det(\Theta_k) - tr(S_k \Theta_k)] - P(\{\Theta_k : k = 1, 2, \ldots, K\}) \right),$$

where K denotes the total number of conditions, n_k denotes the sample size at condition k, and S_k denotes the empirical covariance matrix for the samples collected at condition k. The penalty function $P(\{\Theta_k : k = 1, 2, \ldots, K\})$ is either a fused Lasso penalty

$$P(\{\Theta_k : k = 1, 2, \ldots, K\}) = \lambda_1 \sum_{k=1}^{K} \sum_{i \neq j} |\theta_{k,ij}| + \lambda_2 \sum_{k<k'} \sum_{i,j} |\theta_{k,ij} - \theta_{k',ij}|, \qquad (7.1)$$

or a group Lasso penalty,

$$P(\{\Theta_k : k = 1, 2, \ldots, K\}) = \lambda_1 \sum_{k=1}^{K} \sum_{i \neq j} |\theta_{k,ij}| + \lambda_2 \sum_{i \neq j} \left(\sum_{k=1}^{K} \theta_{k,ij}^2 \right)^{1/2}, \qquad (7.2)$$

where $\theta_{k,ij}$ denotes the $(i, j)^{th}$ element of $\Theta_k \in \mathbb{R}^{p \times p}$, p denotes the dimension of the data, and λ_1 and λ_2 are regularization parameters.

Chun et al. [28] and Guo et al. [57] proposed to work with some penalty functions that regularize the common and condition-specific structures of the graphical models in a hierarchical way. More precisely, Guo et al. [57] proposed to decompose $\theta_{k,ij}$ as

$$\theta_{k,ij} = \omega_{ij} \gamma_{k,ij}, \quad \forall i \neq j,$$

where the component $\omega_{ij} \geq 0$ controls the common structure of the graphical models and the component $\gamma_{k,ij}$ controls their condition-specific structures and suggested the penalty function

$$P(\{\Theta_k : k = 1, 2, \ldots, K\}) = \lambda_1 \sum_{i \neq j} \omega_{ij} + \lambda_2 \sum_{i \neq j} \sum_{k=1}^{K} |\gamma_{k,ij}|. \qquad (7.3)$$

Further, they show that, for computational convenience, the penalty function (7.3) can be replaced by

$$P(\{\Theta_k : k = 1, 2, \ldots, K\}) = \lambda \sum_{i \neq j} g \left(\sum_{k=1}^{K} |\theta_{k,ij}| \right). \qquad (7.4)$$

for some regularization parameter λ and the square root function $g(z) = \sqrt{z}$. Chun et al. [28] generalized the choice for the function $g(\cdot)$.

When K is fairly large, say $K \geq 50$, and the data form a temporal series [134, 203], proposed to modeled them by a high-dimensional time series, where the covariance matrices are regularized to change smoothly over time.

The Bayesian methods enhance the similarity between neighboring graphical models by imposing an appropriate prior on their graph structures directly. For example, Peterson et al. [131] proposed to work with a Markov random field prior. More precisely, they assumed that each edge $l \in \{(i, j) : 1 \leq i < j \leq p\}$ is *a priori* independent of other edges and subject to the prior

$$p(e_l | v_l, R) = C(v_l, R)^{-1} \exp(v_l \mathbf{1}^T e_l + e_l^T R e_l), \tag{7.5}$$

where $e_l = (e_l^{(1)}, e_l^{(2)}, \ldots, e_l^{(K)})$ is a K-binary vector indicating the status of edge l in the K graphs, v_l is an edge-specific parameter, $R \in \mathbb{R}^{K \times K}$ specifies the pairwise relatedness of the K graphs, and $C(v_l, R)$ is the normalizing constant of the distribution. Similarly, Lin et al. [102] imposed a one-dimensional Ising model prior on e_l, which can be viewed as a special case of equation (7.5). The key issue with the Bayesian methods is the computation, which involves inversion of the matrices of size $O(p \times p)$ at each iteration and, therefore, can be very slow when p is large.

FHBIA is very different from the Bayesian methods [102, 131] in that instead of modeling the original data, it models the ψ-scores calculated (as in Chapter 2) for each edge under different conditions. As a result, it avoids inversion of the covariance matrices in iterations and thus can be very efficient for high-dimensional problems. Moreover, it has a very appealing numerical performance.

7.3 Fast Hybrid Bayesian Integrative Analysis

The FHBIA method consists of a few steps: It first applies the ψ-learning method to transform the original data to edgewise ψ-scores under each condition; then it models the ψ-scores in a Bayesian method, based on which it clusters the edges for their status and employs a meta-analysis method to calculate integrated ψ-scores across different conditions; and finally, it performs a multiple hypothesis test to jointly determine the edges for all graphs based on the integrated ψ-scores. Figure 7.1 shows the diagram of the method with the respective steps being described in the remaining part of this section.

7.3.1 ψ-Score Transformation

Suppose that a dataset $\mathcal{X} = \{\mathcal{X}^{(k)} : k = 1, 2, \ldots, K\}$ has been collected under K distinct conditions, where $\mathcal{X}^{(k)} \in \mathbb{R}^{n_k \times p}$ denotes n_k observations collected under condition k, the observations are independent and identically distributed according to a p-dimensional multivariate Gaussian distribution $\mathcal{N}_p(\mu_k, \Sigma_k)$, and μ_k and Σ_k denote the mean and covariance matrix of the distribution, respectively. Note that n_k can be different for different conditions.

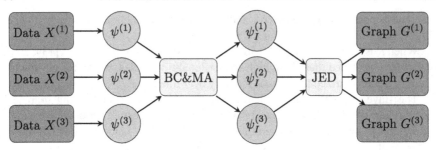

Figure 7.1 *Diagram of the FHBIA method: the data are first transformed to edgewise ψ-scores through ψ-score transformation, the ψ-scores are then processed through Bayesian clustering and meta-analysis (BC&MA) to get integrated ψ-scores (denoted by $\psi_I^{(i)}$'s), and the integrated ψ-scores are further processed through joint edge detection (JED) to get final graph estimates.*

In this step, the ψ-learning algorithm (Algorithm 2.1) is applied to each dataset $\mathcal{X}^{(k)}$ separately, which produces edgewise ψ-partial correlation coefficients for the graphical model under each condition k, and then convert the ψ-partial correlation coefficients to ψ-scores through Fisher's z-transformation and the probit transformation. In what follows, we use $(\psi_l^{(k)})$ for $l \in \{(i,j) : 1 \le i < j \le p\}$ to denote the ψ-partial correlation coefficients calculated with the dataset $\mathcal{X}^{(k)}$ and use $(\Psi_l^{(k)})$ to denote the transformed ψ-scores. Mathematically, we have

$$z_l^{(k)} = \frac{1}{2} \log \left(\frac{1 + \psi_l^{(k)}}{1 + \psi_l^{(k)}} \right),$$

$$\Psi_l^{(k)} = \Phi^{-1} \left(2\Phi \left(\sqrt{n_k - |S_l^{(k)}| - 3} |z_l^{(k)}| \right) - 1 \right), \tag{7.6}$$

where $S_l^{(k)}$ denotes the conditioning set used in the calculation of $\psi_l^{(k)}$ by the ψ-learning algorithm, and $\Phi(\cdot)$ denotes the cumulative distribution function of the standard Gaussian distribution. It is easy to see that $\Psi_l^{(k)} \sim N(0,1)$ if, under condition k, the two variables corresponding to edge l are conditionally independent given all other variables.

7.3.2   *Bayesian Clustering and Meta-Analysis*

To model the ψ-scores, Jia et al. [77] introduced $e_l = (e_l^{(1)}, e_l^{(2)}, \ldots, e_l^{(K)})$ as a vector of latent variables, which indicates the status of edge l in the K underlying true graphs. They assumed that, conditioned on $e_l^{(k)}$, $\Psi_l^{(k)}$'s are mutually independent and follow a two-component mixture Gaussian distribution with the density given by

$$f(\Psi_l^{(k)} | e_l^{(k)}) = [\phi(\Psi_l^{(k)} | \mu_{l0}, \sigma_{l0}^2)]^{1 - e_l^{(k)}} [\phi(\Psi_l^{(k)} | \mu_{l1}, \sigma_{l1}^2)]^{e_l^{(k)}}, \tag{7.7}$$

for $l = 1, 2, \ldots, N$ and $k = 1, 2, \ldots, K$, where $\phi(\cdot|\mu, \sigma^2)$ denotes the density of the Gaussian distribution with mean μ and variance σ^2. Considering the meaning of ψ-scores, one can simply restrict μ_{l0} to 0; however, as shown below, this general treatment of μ_{l0} does not cause much difficulty in computation. Also, the model (7.7) implicitly assumes that the ψ-scores $\{\Psi_l^{(k)} : e_l^{(k)} = 1, k = 1, 2, \ldots, K\}$ follow the same distribution. This assumption can be further relaxed by letting $\{\Psi_l^{(k)} : e_l^{(k)} = 1, k = 1, 2, \ldots, K\}$ follow a mixture Gaussian distribution. Refer to the supplementary material of Jia et al. [77] for the detail, where a three-component mixture Gaussian distribution is considered for the ψ-scores.

Let $\Psi_l = (\Psi_l^{(1)}, \ldots, \Psi_l^{(K)})$. Conditioned on e_l, the joint likelihood function of Ψ_l is given by

$$p(\Psi_l | e_l, \mu_{l0}, \sigma_{l0}^2, \mu_{l1}, \sigma_{l1}^2) = \prod_{\{k : e_l^{(k)} = 0\}} \phi(\Psi_l^{(k)} | \mu_{l0}, \sigma_{l0}^2) \prod_{\{k : e_l^{(k)} = 1\}} \phi(\Psi_l^{(k)} | \mu_{l1}, \sigma_{l1}^2).$$

(7.8)

Therefore, e_l can be inferred under the Bayesian framework by specifying appropriate prior distributions for e_l and other parameters. For example, one can let e_l be subject to a Markov random field prior as in [102, 131], then e_l can be simulated from its posterior using a Markov chain Monte Carlo (MCMC) algorithm, see e.g., [97].

For computational simplicity, Jia et al. [77] let e_l be subject to a Bernoulli process prior with a low probability for edge status alternation. As a result, the marginal posterior of e_l admits a closed form as shown later. Jia et al. [77] considered two types of prior distribution for e_l, temporal prior and spatial prior. The former applies to the case that the graphs evolve along with time, and the latter applies to the case that the graphs evolve independently from a common ancestor graph.

7.3.2.1 Temporal Prior

The temporal prior is given by

$$p(e_l | q) = q^{\sum_{i=1}^{K-1} c_l^{(i)}} (1 - q)^{\sum_{i=1}^{K-1} (1 - c_l^{(i)})},$$

(7.9)

where $c_l^{(i)} = |e_l^{(i+1)} - e_l^{(i)}|$, and q represents the probability of edge status alternation and is subject to a beta distribution:

$$q \sim \text{Beta}(a_1, b_1),$$

(7.10)

where the hyperparameters a_1 and b_1 pre-specified such that q has a small mean value.

Let μ_{l0} and μ_{l1} be subject to an improper uniform distribution, i.e., $\pi(\mu_{l0}) \propto 1$ and $\pi(\mu_{l1}) \propto 1$; and let σ_{l0}^2 and σ_{l1}^2 be subject to an inverse-gamma distribution, i.e.,

$$\sigma_{l0}^2, \sigma_{l1}^2 \sim IG(a_2, b_2),$$

(7.11)

where a_2 and b_2 are pre-specified constants. Then the joint posterior distribution of $(e_l, \mu_{l0}, \sigma_{l0}^2, \mu_{l1}, \sigma_{l1}^2, q)$ is given by

$$\pi(e_l, \mu_{l0}, \sigma_{l0}^2, \mu_{l1}, \sigma_{l1}^2, q | \Psi_l) \propto p(\Psi_l | e_l, \mu_{l0}, \mu_{l1}, \sigma_{l0}^2, \sigma_{l1}^2) \pi(\mu_{l0}, \sigma_{l0}^2, \mu_{l1}, \sigma_{l1}^2)$$
$$\times \pi(e_l | q) \pi(q).$$

Let $n_{l0} = \#\{k : e_l^{(k)} = 0\}$, $n_{l1} = \#\{k : e_l^{(k)} = 1\}$, $k_1 = \sum_{i=1}^{K-1} c_l^{(i)}$ and $k_2 = K - 1 - k_1$. Integrating out the parameters μ_{l0}, σ_{l0}^2, μ_{l1}, σ_{l1}^2 and q from $\pi(e_l, \mu_{l0}, \sigma_{l0}^2, \mu_{l1}, \sigma_{l1}^2, q | \Psi_l)$ leads to the marginal posterior distribution:

$$\pi(e_l | \Psi_l) \propto (H) \times (I) \times (J), \tag{7.12}$$

where the terms (H), (I), and (J) are given as follows:

$$(H) = \frac{\Gamma(a_1 + k_1)\Gamma(b_1 + k_2)}{\Gamma(a_1 + k_1 + b_1 + k_2)},$$

$$(I) = \frac{1}{\sqrt{n_{l0}}} \left(\frac{1}{\sqrt{2\pi}}\right)^{n_{l0}} \Gamma\left(\frac{n_{l0} - 1}{2} + a_2\right)$$
$$\times \left(\frac{1}{2} \sum_{\{k:e_l^{(k)}=0\}} (\Psi_l^{(k)})^2 - \frac{(\sum_{\{k:e_l^{(k)}=0\}} \Psi_l^{(k)})^2}{2n_{l0}} + b_2\right)^{-\left(\frac{n_{l0}-1}{2} + a_2\right)},$$

$$(J) = \frac{1}{\sqrt{n_{l1}}} \left(\frac{1}{\sqrt{2\pi}}\right)^{n_{l1}} \Gamma\left(\frac{n_{l1} - 1}{2} + a_2\right)$$
$$\times \left(\frac{1}{2} \sum_{\{k:e_l^{(k)}=1\}} (\Psi_l^{(k)})^2 - \frac{(\sum_{\{k:e_l^{(k)}=1\}} \Psi_l^{(k)})^2}{2n_{l1}} + b_2\right)^{-\left(\frac{n_{l1}-1}{2} + a_2\right)}.$$

In the derivation, Ψ_l is treated as observed data and $n_{l0} > 0$ and $n_{l1} > 0$ are assumed. It is easy to show that $\pi(e_l | \Psi_l) \propto (H) \times (J)$ if $n_{l0} = 0$ and $n_{l1} > 0$, and $\pi(e_l | \Psi_l) \propto (H) \times (I)$ if $n_{l0} > 0$ and $n_{l1} = 0$.

Let $\{e_{ld} : d = 1, 2, \ldots, 2^K\}$ denote the set of all possible configurations of e_l. When K is small, $\pi(e_l | \Psi_l)$ can be evaluated exhaustively over the set $\{e_{ld} : d = 1, 2, \ldots, 2^K\}$. Let $\pi_{ld} = \pi(e_{ld} | \Psi_l)$, and let $\bar{\Psi}_{ld} = (\bar{\Psi}_{ld}^{(1)}, \ldots, \bar{\Psi}_{ld}^{(K)})$ denote the integrated Ψ-scores given the configuration $e_{ld} = (e_{ld}^{(1)}, e_{ld}^{(2)}, \ldots, e_{ld}^{(K)})$. By Stouffer's meta-analysis method [157],

$$\bar{\Psi}_{ld}^{(k)} = \begin{cases} \dfrac{\sum_{\{i:e_{ld}^{(i)}=0\}} w_i \Psi_l^{(i)}}{\sqrt{\sum_{\{i:e_{ld}^{(i)}=0\}} w_i^2}}, & \text{if } e_{ld}^{(k)} = 0, \\[3ex] \dfrac{\sum_{\{i:e_{ld}^{(i)}=1\}} w_i \Psi_l^{(i)}}{\sqrt{\sum_{\{i:e_{ld}^{(i)}=1\}} w_i^2}}, & \text{if } e_{ld}^{(k)} = 1, \end{cases} \tag{7.13}$$

where w_i denotes the weight assigned to condition i, depending on the size n_i and quality of the dataset $\mathcal{X}^{(i)}$. For example, for simplicity, one can simply set $w_i = 1$ for $i = 1, 2, \ldots, K$. Then the integrated ψ-score is given by

$$\hat{\Psi}_l^{(k)} = \sum_{d=1}^{2^K} \pi_{ld} \bar{\Psi}_{ld}^{(k)}, \quad k = 1, 2, \ldots, K, \quad l \in \{(i, j) : 1 \le i < j \le p\}. \tag{7.14}$$

When K is large, π_{ld} can be estimated through MCMC, say, running evolutionary Monte Carlo (EMC) algorithm [99] for a few hundred iterations, which is particularly efficient for sampling from a discrete sample space. Alternatively, as suggested by Sun et al. [161], π_{ld} can be estimated using the stochastic EM algorithm [20, 123]. Since the estimation can be processed in parallel for different edges, the computation should not be a big concern in this case.

7.3.2.2 Spatial Prior

The spatial prior is given by

$$p(e_l|q) = q^{\sum_{i=1}^{K} c_l^{*(i)}} (1 - q)^{\sum_{i=1}^{K} (1 - c_l^{*(i)})}, \tag{7.15}$$

where $c_l^{*(i)} = |e_l^{(i)} - e_l^{\text{mod}}|$, e_l^{mod} represents the most common status of edge l across all K graphs, and the parameter q is also subject to the prior (equation 7.10). With this prior, $\pi(e_l|\Psi_l)$ has also an analytic form as given in equation (7.12) but with $k_1 = \sum_{i=1}^{K} c_l^{*(i)}$ and $k_2 = K - k_1$.

7.3.3 Joint Edge Detection

The structures of multiple Gaussian graphical models can be jointly determined by performing a multiple hypothesis test (say, e.g., [95]) on the integrated ψ-scores (equation 7.14). The multiple hypothesis test groups the integrated ψ-scores into two classes: presence of edges and absence of edges. The threshold value used in grouping is determined by α_2, a pre-specified significance level for the multiple hypothesis test [156]. Refer to Appendix A.7 for a brief description of the empirical Bayesian multiple hypothesis testing method [95].

7.3.4 Parameter Setting

FHBIA contains a few parameters, including $\alpha_1, \alpha_2, a_1, b_1, a_2$ and b_2. The parameter α_1 is used in the step of ψ-score transformation, which specifies the significance level of the multiple hypothesis test in correlation screening (refer to Algorithm 2.1). As suggested before, α_1 can be set to a slightly large value such as 0.05, 0.1, or 0.2. The parameter α_2 is defined in Section 7.3.3, which controls the sparsity level of the graph estimates. The prior hyperparameters (a_1, b_1) are defined in equation (7.10), and the prior hyperparameters (a_2, b_2) are defined in equation (7.11). To enhance the structure smoothness of the graphs, Jia et al. [77] suggested to set $(a_1, b_1) = (1, 10)$ as the default, which leads to a small prior probability for neighboring edge alternation.

Motivated by the observation that $\Psi_l^{(k)} \sim N(0,1)$ when $e_l^{(k)} = 0$ is true, Jia et al. [77] suggested to set $(a_2, b_2) = (1,1)$ as the default.

7.4 Simulation Studies

This section compares the performance of FHBIA with some existing methods under both scenarios of temporal and spatial priors.

7.4.1 Scenario of Temporal Prior

In this study, we considered five types of network structures, namely, band (AR(2)), cluster, random, scale-free, and hub. Refer to [200] or [76] for how to generate networks with these types of network structures. For each type of network structure, we set $K = 4$, $n_1 = n_2 = \cdots = n_K = 100$ and $p = 200$. Under each condition $k \in \{1, 2, \ldots, K\}$, we generated ten datasets from $\mathcal{N}_p(0, \Theta_k^{-1})$.

To construct Θ_k based on Θ_{k-1} for $k = 2, 3, \ldots, K$, a random edge deletion-addition procedure [102, 131] is used: (i) set $\Theta_k = \Theta_{k-1}$; (ii) randomly select 5% nonzero elements in Θ_k and set them to 0; (iii) randomly select the same number of zero elements in the modified Θ_k and replace them by random draws from the uniform distribution $\frac{1}{2}\text{Unif}[-0.4, -0.2] + \frac{1}{2}\text{Unif}[0.2, 0.4]$; and (iv) modify the diagonal elements of Θ_k to ensure Θ_k to be positive definite. See Appendix A.8.5 for the code used in the simulation study.

Table 7.1 compares averaged AUCs (areas under the precision-recall curves) produced by FHBIA, fused graphical Lasso (FGL), and group graphical Lasso (GGL)[1] [34], and separated ψ-learning on this example. For FGL and GGL, we tried a large number of settings of (λ_1, λ_2). In particular, for each value of λ_2, we tried 20 values of λ_1 which are equally spaced between 0.025 and 0.725 such that its recall values vary in a wide range and the corresponding precision-recall curve can be well approximated by the 20 points of (recall, precision); then we carefully selected the values of λ_2 for each type of network structures such that the maximum AUCs of the method can be approximately obtained. In the separated ψ-learning method, the graphs were constructed separately under each condition. We included it for the comparison in order to demonstrate the importance of data integration for high-dimensional graphical modeling. The comparison shows that FHBIA significantly outperforms FGL and GGL for the structures AR(2), cluster, random, and scale-free, and performs equally well as FGL and GGL for the structure hub.

Figure 7.2 illustrates the performance of the FHBIA method for the networks with the random structure. The upper panel shows how the network evolves along with condition changes. The lower panel shows the estimated networks, which clearly indicate the effect of network integration by noting the structural similarities between neighboring networks.

---

[1]The implementations of FGL and GGL are available at the R package *JGL* [33].

Table 7.1 *Averaged AUCs produced by different methods for ten datasets simulated under the scenario of temporal priors, where the number in the parentheses denotes the standard deviation of the averaged AUC.*

Structure	Measure	GGL	FGL	ψ-Learning	FHBIA
AR(2)	ave-AUC	0.6909	0.7856	0.6780	0.8686
	SD	(0.0036)	(0.0034)	(0.0036)	(0.0025)
Cluster	ave-AUC	0.6490	0.7397	0.5034	0.8452
	SD	(0.0105)	(0.0102)	(0.0082)	(0.0053)
Random	ave-AUC	0.8540	0.9129	0.5998	0.9335
	SD	(0.0072)	(0.0045)	(0.0162)	(0.0030)
Scale free	ave-AUC	0.8680	0.9266	0.6130	0.9429
	SD	(0.0094)	(0.0029)	(0.0072)	(0.0032)
Hub	ave-AUC	0.9441	0.9504	0.7889	0.9477
	SD	(0.0015)	(0.0015)	(0.0061)	(0.0027)

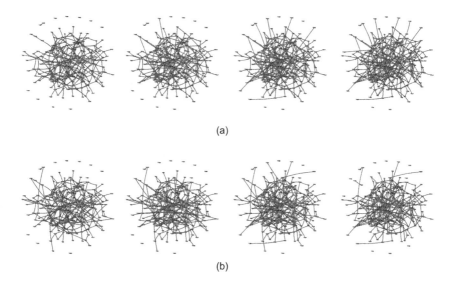

(a)

(b)

Figure 7.2 *Joint estimation of multiple networks: the upper panel (a) shows the network structures under four distinct conditions, and the lower panel (b) shows the estimated networks by FHBIA.*

7.4.2 Scenario of Spatial Prior

We considered again the five types of network structures. For each of them, we first generated the initial precision matrix Θ_0 as in Section 7.4.1. Then, we generated Θ_k's for $k = 1, 2, \ldots, K$ by evolving from Θ_0 independently according to the random edge deletion-addition procedure described in Section 7.4.1. In this study, we set $K = 4$,

Table 7.2 *Averaged AUCs produced by different methods for ten datasets simulated under the scenario of spatial priors, where the number in the parentheses denotes the standard deviation of the averaged AUC.*

Structure	Measure	GGL	FGL	ψ-Learning	FHBIA
AR(2)	ave-AUC	0.6922	0.7949	0.6635	0.8580
	SD	(0.0066)	(0.0064)	(0.0072)	(0.0052)
Cluster	ave-AUC	0.6407	0.7297	0.5098	0.8497
	SD	(0.0058)	(0.0068)	(0.0072)	(0.0046)
Random	ave-AUC	0.8445	0.9054	0.6052	0.9408
	SD	(0.0095)	(0.0049)	(0.0154)	(0.0027)
Scale free	ave-AUC	0.8548	0.9179	0.6193	0.9518
	SD	(0.0086)	(0.0027)	(0.0086)	(0.0018)
Hub	ave-AUC	0.9581	0.9536	0.8074	0.9593
	SD	(0.0016)	(0.0012)	(0.0045)	(0.0009)

$p = 200$, and $n_1 = n_2 = \cdots = n_K = 100$, and generated 10 datasets for each type of network structure.

The FHBIA, FGL, GGL, and separated ψ-learning methods were applied to the simulated data. The results are summarized in Table 7.2. Similar to the temporal case, the comparison in this case indicates again the superiority of the FHBIA method over the others.

We note that other than FGL and GGL, the graphical EM [182] method can also be used for joint estimation of multiple Gaussian graphical models under the scenario of spatial prior. The graphical EM method decomposes each graph into two graphical layers, one for the common structure across different conditions and the other for condition-specific structures, and then estimates the concentration matrices of the graphs using the EM algorithm [35] by treating the common structure as latent variables. However, the numerical results in Jia et al. [77] indicate that it generally performed less favorably compared to FGL and FHBIA.

7.5 Application

Jia et al. [77] applied the FHBIA method to an mRNA gene expression dataset collected by the TEDDY (abbreviation for *The Environmental Determinants of Diabetes in the Young*) group. Their results are briefly reviewed as follows.

The dataset consists of 21,285 genes and 742 samples (half cases and half control) collected at multiple time points from 313 subjects. For each subject, the dataset also contains some demographical variables such as age (data collection time), gender, race, season of birth, etc. In pre-processing the dataset, Jia et al. [77] filtered out inactive genes (across the case-control groups) and removed small group samples (in ages). After pre-processing, they got 582 samples (half cases and half control), each sample consisting of 572 genes. Those samples from nine groups in ages with the

Case

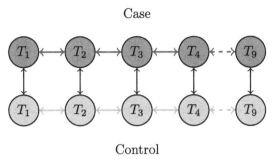

Control

Figure 7.3 *Data integration scheme for TEDDY data, where the arrows indicate that data integration is to be done within cases, within control, and across the case and control groups.*

group sizes ranging from 42 to 98, and each group consists of half case samples and half control samples. Figure 7.3 shows the structure of the pre-processed data, where the arrows indicate that the data integration is to be done across nine time points as well as the case and control groups.

Jia et al. [77] calculated the ψ-scores using the method described in Section 2.6.3, which adjusts the effect of demographical variables. Then they integrated the ψ-scores in two steps: they first made the integration across nine time points for the case and control groups separately, and then made the integration across the case and control groups for each time point. Finally, they applied the empirical Bayesian multiple hypothesis test [95] to the integrated ψ-scores to determine the structures of eighteen networks. Through this study, they identified some hub genes for the disease and found some interesting changes in the gene regulatory network structure along with disease progression. Refer to [77] for detail.

7.6   Consistency

As shown in Section 2.7, the multiple hypothesis test based on ψ-scores produces a consistent estimate for the Gaussian graphical model with data observed under a single condition. To accommodate the extension from single condition to multiple conditions, Jia et al. [77] modified the assumptions used in Section 2.7 and added an assumption about the number of conditions K. Their new set of assumptions is given as follows.

Suppose that a dataset $\mathcal{X} = \{\mathcal{X}^{(k)} : k = 1, 2, \ldots, K\}$ has been collected under K distinct conditions, where $\mathcal{X}^{(k)} \in \mathbb{R}^{n_k \times p}$ denotes the dataset under condition k, n_k is the same size, and the observations are independent and identically distributed according to a p-dimensional multivariate Gaussian distribution $\mathcal{P}_k^{(n_k)}$. Without loss of generality, we assume that $n_1 = n_2 = \cdots = n_K = n$. Let G_k denote the true graph under condition k, and let $E_k = \{(i, j) : \rho_{ij|V \setminus \{i,j\}}^{(k)} \neq 0, \ i, j = 1, \ldots, p\}$ denotes the edge set of graph G_k, where $\rho_{ij|V \setminus \{i,j\}}^{(k)}$ denotes the partial correlation coefficient of variable i and variable j under condition k. To indicate that the dimension p and the number of conditions K can increase with the sample size n, they are re-notated by p_n and K_n, respectively.

Assumption 7.1 *The distribution $\mathcal{P}_k^{(n)}$ is multivariate Gaussian and satisfies the Markov property and faithfulness condition with respect to the undirected graph G_k for each $k = 1, 2, \ldots, K_n$ and $n \in \mathbb{N}$.*

Assumption 7.2 *The dimension $p_n = O(\exp(n^\delta))$ for some constant $0 \leq \delta < 1$, and is the same under each condition.*

Assumption 7.3 *The correlation coefficients satisfy*

$$\min\left\{|r_{ij}^{(k)}| : r_{ij}^{(k)} \neq 0, \ i,j = 1,2,\ldots,p_n, \ i \neq j, k = 1,2\ldots,K_n\right\} \geq c_0 n^{-\kappa}, \quad (7.16)$$

for some constants $c_0 > 0$ and $0 < \kappa < (1-\delta)/2$, and

$$\max\left\{|r_{ij}^{(k)}| : i,j = 1,\ldots,p_n, i \neq j, k = 1,2,\ldots,K_n\right\} \leq M_r < 1, \quad (7.17)$$

for some constant $0 < M_r < 1$, where $r_{ij}^{(k)}$ denotes the correlation coefficient of variable i and variable j under condition k.

Assumption 7.4 *There exist constants $c_2 > 0$ and $0 \leq \tau < 1 - 2\kappa'$ such that $\max_k \lambda_{\max}(\Sigma_k^{(n)}) \leq c_2 n^\tau$, where $\Sigma_k^{(n)}$ denotes the covariance matrix of $\mathcal{P}_k^{(n)}$, and $\lambda_{\max}(\Sigma_k^{(n)})$ is the largest eigenvalue of $\Sigma_k^{(n)}$.*

Assumption 7.5 *The ψ-partial correlation coefficients satisfy $\inf\{\psi_{ij}^{(k)} : \psi_{ij}^{(k)} \neq 0, \ 0 < |S_{ij}^{(k)}| \leq q_n, \ i,j = 1,\ldots,p_n, \ i \neq j, \ k = 1,2,\ldots,K_n\} \geq c_3 n^{-d}$ for some constant $c_3 > 0$, where $\psi_{ij}^{(k)}$ denotes the ψ-partial correlation coefficient of variable i and variable j under condition k, $q_n = O(n^{2\kappa'+\tau})$, $0 < d < (1-\delta)/2$, and $S_{ij}^{(k)}$ is the neighboring set used in calculation of $\psi_{ij}^{(k)}$. In addition,*

$$\sup\left\{\psi_{ij}^{(k)}; \ 0 < |S_{ij}^{(k)}| \leq q_n, \ i,j = 1,\ldots,p_n, \ i \neq j, \ k = 1,2,\ldots,K_n\right\} \leq M_\psi < 1,$$

for some constant $0 < M_\psi < 1$.

Assumption 7.6 *$K_n = O(n^{\delta+2d+\varepsilon-1})$ for some constant $\varepsilon > 0$ such that $\delta + 2d + \varepsilon - 1 \geq 0$, where δ is as defined in Assumption 7.2 and d is as defined in Assumption 7.5.*

Assumption 7.6 is mild. For example, if one sets $\varepsilon = 1 - \delta - 2d$, then $K_n = O(1)$. This is consistent with the numerical performance of FHBIA that it can work well with a small value of K_n. Under the above assumptions, Jia et al. [77] proved the consistency of the FHBIA method. Refer to the supplementary material of Jia et al. [77] for the proof.

Theorem 7.1 *(Theorem 2.1; [77]) Suppose Assumptions 7.1–7.6 hold. Then*

$$P\left[\widehat{E}_{k,\zeta_n}^{(n)} = E_k, k = 1,2,\ldots,K_n\right] \geq 1 - o(1), \quad \text{as } n \to \infty,$$

COMPUTATIONAL COMPLEXITY 69

where $\widehat{E}_{k,\zeta_n}^{(n)}$ denotes the FHBIA estimator of E_k, and ζ_n denotes a threshold value of integrated ψ-scores based on which the edges are determined for all K_n graphs.

Jia et al. [77] particularly pointed out that as implied by their proof, the meta-analysis step indeed improves the power of the FHBIA method in the identification of the edge sets $E_1, E_2, \ldots, E_{K_n}$. Refer to [77] for more discussions on this issue.

7.7 Computational Complexity

FHBIA is computationally efficient. It has a computational complexity of $O(p^2 2^K)$, and it can be further accelerated by the stochastic EM algorithm as suggested by Sun et al. [161] when K is large. By directly modeling the edgewise ψ-scores, the FHBIA method avoids the inversion of high-dimensional precision/covariance matrices that other Bayesian methods, e.g., [102, 131], suffer from. Moreover, like the ψ-learning method, many steps of the method can be easily paralleled, e.g., the edge-wise Bayesian clustering and meta-analysis step, the correlation coefficient calculation step, and the ψ-score calculation step. Therefore, the execution of the method can also be further accelerated with a parallel architecture.

7.8 Problems

1. Provide the detail for the derivation of the marginal posterior (equation 7.12).

2. Prove Theorem 7.1.

3. Redo the simulation studies in Section 7.4.

4. Extend the FHBIA method to joint estimation of multiple mixed graphical models [75] and perform a simulation study to test its performance.

5. Derive the formulas in Section 7.3.2 for the case of three-component mixture Gaussian distribution.

6. Perform a simulation study with a large number of conditions and estimate π_{ld}'s in equation (7.14) using evolutionary Monte Carlo [99].

Chapter 8

Nonlinear and Non-Gaussian Graphical Models

This chapter introduces a double regression method for learning high-dimensional nonlinear and non-Gaussian graphical models based on the work [101]. The method is very general, which, in theory, can be applied to any type of data provided that the required variable selection/screening procedure and conditional independence test are available.

8.1 Introduction

In previous chapters, we have considered both cases where the data are Gaussian and non-Gaussian. The former is covered by Chapters 2–4, while the latter is covered by Chapters 5 and 6. Chapter 5 suggests transformations, including data-continuized transformation [74] and data-Gaussianized transformation [103], which make graphical modeling possible for non-Gaussian continuous data and some types of discrete data such as Poisson and negative binomial. Chapter 6 deals with mixed data particularly the case of Gaussian and multinomial data. In both chapters, the variables are assumed to be pairwise and linearly dependent, although the work of Cheng et al. [23] shows that including some higher-order interaction terms in the model is helpful for revealing relationships between different variables. In this chapter, we consider graphical modeling under the most general scenario that the variables can be non-Gaussian and the interactions between different variables can be nonlinear and of high orders.

As mentioned in Chapter 1, there are only very few works on nonlinear and non-Gaussian graphical modeling in the literature, which include [101, 191, 202]. In what follows, we first give a brief review for Yu et al. [191] and Zheng et al. [202] and then give a more detailed description for Liang and Liang [101].

8.2 Related Works

Zheng et al. [202] extended the structural equation modeling method [201], see Section 2.2.4, to nonlinear models defined in Sobolev space [170]. Let f_j denote the mean function fitted for the nonlinear regression $X_j \sim X_{\backslash j}$, where $X_{\backslash j} = \{X_1, X_2, \ldots, X_{j-1}, X_{j+1}, \ldots, X_p\}$. Let $\partial_i f_j$ denote its partial derivative with respect to

DOI: 10.1201/9780429061189-8

X_i. Zheng et al. [202] proposed to replace the coefficient $\beta_{j,i}$ in the linear regression (equation 2.6) by $\|\partial_i f_j\|_2$ based on the mathematical fact that

$$f_j \text{ is independent of } X_i \iff \|\partial_i f_j\|_2 = 0.$$

Suppose that f_j is modeled by a $(L+1)$-layer neural network with parameters denoted by $\theta_j = (A_j^{(1)}, \ldots, A_j^{(L)})$, where $A_j^{(l)}$ denotes the parameters for layer l and its ith column denotes the weights from neuron i of layer l to the neurons of layer $l+1$. Let $\theta = (\theta_1, \theta_2, \ldots, \theta_p)$, and let $[W(\theta)]_{ij} = \|i^{th}\text{-column}(A_j^{(1)})\|_2$. Therefore, $\|\partial_i f_j\|_2 = 0$ if $[W(\theta)]_{ij} = 0$. Further, Zheng et al. [202] proposed to learn a nonlinear and non-Gaussian DAG by solving the following constrained optimization problem:

$$\min_{\theta} \frac{1}{n} \sum_{j=1}^{p} \left\{ -l(x_j, NN(x_{\setminus j}; \theta_j)) + \lambda \|vec(A_j^{(1)})\|_1 \right\}, \tag{8.1}$$

$$\text{subject to} \quad h(W(\theta)) = 0,$$

where $l(\cdot, \cdot)$ denotes the log-likelihood function of the neural network, $h(B) = tr(e^{B \circ B}) - p^2$ is the acyclicity constraint as defined in equation (2.6), λ is a regularization parameter, and the L_1-penalty induces a neural network with a sparse input layer. Unfortunately, such a neural network might not be able to provide a consistent estimator for the function f_j, especially when the training sample size n is smaller than the dimension p. In consequence, the DAG cannot be consistently identified by the method. We call this method "DAG-notears" in this chapter.

Yu et al. [191] inferred the structure of the graphical model by learning a weighted adjacency matrix of a DAG with variational autoencoder. Let $X = (X_1, X_2, \ldots, X_p)^T$ denote a p-dimensional random vector, let $Z \in \mathbb{R}^p$ denote a noise vector, and let A denote a weighted adjacency matrix which is strictly upper triangular. By mimicking the linear structural equation of the Gaussian graphical model:

$$X = A^T X + Z,$$

they proposed a deep generative neural network (GNN) model such that

$$\text{Decoder}: \quad X = f_2((I - A^T)^{-1} f_1(Z)),$$
$$\text{Enconder}: \quad Z = f_4((I - A^T) f_3(X)),$$

where f_1, f_2, f_3 and f_4 are parameterized nonlinear functions. In particular, they set f_2 and f_4 to be the identity function, and f_1 and f_2 are approximated by multilayer perceptrons (MLPs). Then they let A be subject to the acyclicity constraint and learn A and the parameters of the two MLPs jointly using the variational method. Like the method DAG-notears, this method lacks theoretical guarantee for the consistency of its DAG estimator. We call this method "DAG-GNN" in this chapter.

8.3   Double Regression Method

The double regression method [101] works based on equation (1.7) directly. That is, for any pair of random variables (X_i, X_j), it first reduces the conditioning set of the conditional independence test through two regressions $X_i \sim X_{V \setminus \{i\}}$ and $X_j \sim X_{V \setminus \{i,j\}}$, where $V = \{1, 2, \ldots, p\}$ is the index set of all variables, and then performs the conditional independence test with the reduced conditioning set. This procedure is summarized as the following algorithm.

Algorithm 8.1 *Double Regression [101]*

i. *For each variable X_i, $i \in V$, perform a variable sure screening procedure for the nonlinear regression*

$$X_i \sim X_{V \setminus \{i\}}, \tag{8.2}$$

and denote the index set of the selected variables by $\hat{S}_i$.

ii. *For each pair of variables (X_i, X_j), $1 \leq j < i \leq p_n$, perform a variable sure screening procedure for the nonlinear regression*

$$X_j \sim X_{V \setminus \{i,j\}}, \tag{8.3}$$

and denote the index set of the selected variables by $\hat{S}_{j \setminus i}$.

iii. *For each pair of variables (X_i, X_j), perform the conditional independence test*

$$X_i \perp\!\!\!\perp X_j | X_{\hat{S}_i \cup \hat{S}_{j \setminus i} \setminus \{j\}}, \tag{8.4}$$

and denote the p-value of the test by q_{ij}.

iv. *Identify the edges of the graphical models by performing a multiple hypothesis test on the individual p-values $\{q_{ij} : 1 \leq j < i \leq p\}$ learned in step (iii).*

This algorithm is general and can be applied to any type of data. However, as mentioned in Section 1.2, the conditioning set $\hat{S}_i \cup \hat{S}_{j \setminus i} \setminus \{j\}$ can be further reduced if more information on the data X is known. For example, it can be much reduced for Gaussian data with linear dependence as described in Chapter 2, and mixed data (from exponential family distributions) with pairwise dependence as described in Chapter 6.

The procedures performed in equations (8.2) and (8.3) are generally required to satisfy the sure screening property, i.e., $S_i \subseteq \hat{S}_i$ and $S_{j \setminus i} \subseteq \hat{S}_{j \setminus i}$ hold almost surely as the sample size $n \to \infty$, where S_i and $S_{j \setminus i}$ are as defined in equation (1.7). There are many procedures satisfying this property, for example, sparse Bayesian neural networks have been shown to have this property [96, 162]. Alternatively, a model-free sure independence screening procedure can be applied here. For example, the Henze-Zirkler sure independence screening (HZ-SIS) procedure [184] can be applied when X_i's are continuous, and an empirical conditional distribution function-based variable screening procedure [32] can be applied when X_i's are categorical.

For the conditional independence test (equation 8.4), Liang and Liang [101] suggested a nonparametric method as the functional form of the dependency between different variables is unknown. There are quite a few nonparametric conditional independence tests available in the literature, including permutation-based tests [12, 39], kernel-based tests [158, 197], classification or regression-based tests [144, 145, 198], and generative adversarial network (GAN)-based tests [9], among others. Refer to [90] for a comprehensive overview.

Under mild conditions, Liang and Liang [101] proved that the double regression method is consistent. We deferred this theory to Section 8.5.

8.4   Simulation Studies

We illustrate the performance of the double regression method using both low- and high-dimensional problems.

8.4.1   A Low-Dimensional Example

This example was modified from an example of Liang and Liang [101], which consisted of 50 datasets simulated from the following nonlinear and non-Gaussian model:

$$
\begin{aligned}
X_1 &\sim \text{Unif}[-1,1], \\
X_2 &= 6\cos(X_1) + \varepsilon_2, \quad \varepsilon_2 \sim \text{Unif}[-1,1], \\
X_3 &= 5\sin(X_1) + X_2 + \varepsilon_3, \quad \varepsilon_3 \sim N(0,1), \\
X_4 &= 5\cos(X_3 X_6) + 3X_3 + 3X_6 + \varepsilon_4, \quad \varepsilon_4 \sim N(0,1), \\
X_5 &= 0.05(X_2 + X_6)^3 + \varepsilon_5, \quad \varepsilon_5 \sim N(0,1), \\
X_6 &\sim \text{Unif}[-1,1], \\
X_7 &= 6\cos(0.2(X_3 + \log(|5X_5| + 1))) + \varepsilon_7, \quad \varepsilon_7 \sim \text{Unif}[-1,1], \\
X_i &\sim N(0,1), \quad i = 8,9,\ldots,p.
\end{aligned}
\tag{8.5}
$$

Each dataset consisted of $n = 200$ observations and $p = 30$ variables which are uniformly or normally distributed.

The double regression method was applied to this example, where HZ-SIS [184] was used in variable sure screening, the generalized covariance measure (GCM) test [145] was used in conditional independence tests, and the adjusted p-values [66] were used for the multiple hypothesis test. Figure 8.1 shows the Markov network recovered by the method for one simulated dataset, where the neighborhood size n_s was set to 5 in variable screening, i.e., $\hat{S}_i$ and $\hat{S}_{j\backslash i}$ contain only the most relevant 5 variables in the respective regression model, and the multiple hypothesis test was conducted at a significance level of $\alpha = 0.05$. It is remarkable that the subnetwork identified for the variables $X_1 - X_7$ is identical to the true one; that is, the double regression method has successfully recovered the parents/children links in the model (8.5) as well as the spouse links induced by the model. See Section 6.3 for how an undirected Markov network can be constructed from a DAG by moralization.

Table 8.1 summarizes the performance of the double regression method by the average AUC values (under the precision-recall curves), where different neighborhood sizes $n_s = 5$, 8, and 10 were tried. The double regression method has been compared to DAG-notears[1] and DAG-GNN[2] in Liang and Liang [101] in a similar example, where each dataset was simulated from the same model (8.5) but consisting of 400 independent random samples. However, even with the doubled sample size, the AUC values produced by DAG-notears and DAG-GNN are still lower than 0.8.

---

[1]The program code is available at https://github.com/xunzheng/notears.
[2]The program code is available at https://github.com/fishmoon1234/DAG-GNN.

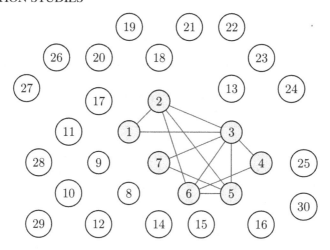

Figure 8.1 *Undirected Markov network recovered by the double regression method for one dataset simulated from the model (8.5): the subnetwork for the variables $X_1 - X_7$ is highlighted, where the red lines indicate parents/children links and the blue lines indicate spouse links.*

For DAG-notears, different options for approximating the regression (equation 8.2) have been tried, including the linear function, multilayer perceptron, and Sobolev basis function. But the resulting AUC values are all lower than 0.8. This comparison indicates the superiority of the double regression method for this example.

Finally, we note that Liang and Liang [101] employed the randomized conditional correlation test (RCoT) [158] for performing nonparametric conditional independence tests, and their results are inferior to those reported in Table 8.1. However, the RCoT test can be much faster than the GCM test used above.

8.4.2 A High-Dimensional Example

This example was also modified from an example of Liang and Liang [101]. It consisted of ten datasets simulated from the following nonlinear and non-Gaussian model:

$$
\begin{aligned}
& X_1 \sim \text{Unif}[-1,1], \quad X_2 = g(X_1) + \varepsilon_2, \quad \varepsilon_2 \sim N(0,1) \\
& X_i = f(X_{i-2}) + g(X_{i-1}) + \varepsilon_i, \quad i = 3,4,\ldots,p,
\end{aligned}
\tag{8.6}
$$

Table 8.1 *Averaged AUC values (over 50 datasets, under precision-recall curves) by the double regression method for discovering the undirected Markov network of the model (8.5), where "SD" represents the standard deviation of the averaged AUC value.*

Neighborhood size	5	8	10
ave-AUC	0.9322	0.9111	0.9003
SD	0.0063	0.0085	0.0097

Table 8.2 *Average AUC values (over ten datasets, under precision-call curves) produced by the double regression method for the model (8.6), where "SD" represents the standard deviation of the average AUC value.*

Neighborhood size	20	30	40	50
ave-AUC	0.7062	0.7202	0.7263	0.7293
SD	0.0035	0.0041	0.0045	0.0033

where $\varepsilon_i \sim \frac{1}{2}N(0,1) + \frac{1}{2}\text{Unif}[-0.5, 0.5]$ for $i = 3, 4, \ldots, p$, and the functions $f(z)$ and $g(z)$ are randomly drawn from the set $\{|z|\cos(z), \tanh(z), \log(|z|+1)\}$ which might vary for different values of $i \in \{2, 3, \ldots, p\}$. In simulations, we set $n = 300$ and $p = 500$ such that this example represents a small-n-large-p problem.

The double regression method was applied to this example as in Section 8.4.1 except that the GCM test was replaced by the RCoT in the step of conditional independence tests. The results were summarized in Table 8.2. The DAG-notears and DAG-GNN methods were not included in comparison for this example, since, as shown in Liang and Liang [101], they essentially failed for small-n-large-p problems.

8.5 Consistency

This section established the consistency of the double regression method under the high-dimensional scenario. Let T_{ij} denote the test statistic in equation (8.4), let p_n denote the dimension of the data, let E_n denote the adjacency matrix of the true network, and let $\hat{E}_n$ denote the estimator of E_n by the double regression method. To study the consistency of $\hat{E}_n$, Liang and Liang [101] made the following assumptions.

Assumption 8.1 *(Markov and faithfulness) The generative distribution of the data is Markovian and faithful with respect to a DAG.*

Assumption 8.2 *(High dimensionality) The dimension p_n increases in a polynomial of the sample size n.*

Assumption 8.3 *(Uniform sure screening property) The variable screening procedure satisfies the uniform sure screening property, i.e., $\min_{1 \leq i \leq p_n} P(S_i \subset \hat{S}_i) \to 1$ and $\min_{1 \leq j < i \leq p_n} P(S_{j\backslash i} \subset \hat{S}_{j\backslash i}) \to 1$ hold as the sample size $n \to \infty$.*

Assumption 8.4 *(Test mean separation) $\min_{i,j}(\mu_{ij,1} - \mu_{ij,0}) > \eta_n$, where $\eta_n = c_0 n^{-\kappa}$ for some constant $c_0 > 0$ and $\kappa > 0$, and $\mu_{ij,0}$ and $\mu_{ij,1}$ denote the mean values of the test statistic T_{ij} under the null (conditionally independent) and alternative (conditionally dependent) hypotheses, respectively.*

Assumption 8.5 *(Tail probability)*

$$\sup_{i,j} P\left(|T_{ij} - \mu_{ij}| > \frac{1}{2}\eta_n\right) = \exp\left\{-O(n^{\delta(\kappa)})\right\},$$

where $\mu_{ij} = E(T_{ij})$ denotes the expectation of T_{ij}, and $\delta(\kappa)$ is positive number depending on κ.

Assumptions 8.1–8.3 are regular, which have been assumed or proved for linear Gaussian models in previous chapters of this book. Assumption 8.4 is like a beta-min condition used in high-dimensional variable selection, see e.g. [37, 174], which ensures the means of the test statistics to be separable under the null and alternative hypotheses. Assumption 8.5 constrains the tail probabilities of the distributions of the test statistics. For example, if T_{ij} is an average-type test statistic and follows a sub-Gaussian distribution, then the assumption can be satisfied by the concentration theorem. Under these assumptions, Liang and Liang [101] established the following theorem:

Theorem 8.1 *Suppose Assumptions 8.1–8.5 hold. For each pair of variables (X_i, X_j), let $\mu_{ij,0} + \eta_n/2$ denote the critical value of the conditional independence test (8.4). Then $\hat{E}_n$ is consistent, i.e.,*

$$P\{an\ error\ occurs\ in\ \hat{E}_n\} \to 0, \quad as\ n \to \infty.$$

8.6 Computational Complexity

A plain implementation of the double regression method can have a computational complexity of $O(p_n^2(R(n, p_n) + T(n, p_n)))$, where $R(n, p_n)$ and $T(n, p_n)$ denote, respectively, the computational complexities of a single regression task and a single conditional independence test in Algorithm 8.1. However, this complexity can be much reduced, particularly when $R(n, p_n)$ is high. For example, Liang and Liang [101] prescribed a Markov blanket-based algorithm, which works in a similar way to Algorithm 6.1 and avoids performing regressions in step (ii) of Algorithm 8.1. As a result, the computational complexity of the algorithm can be reduced to $O(p_n R(n, p_n) + p_n^2 T(n, p_n))$.

The term $O(p_n^2 T(n, p_n))$ can be further reduced based on the sure screening property of the regression procedure performed in step (i). Suppose that there are $O(n/\log(n))$ variables selected for each regression in step (i) and the conditional independence tests are performed only for the pairs of variables covered by the Markov blankets. Then the computational complexity of the double regression method can be reduced to $O(p_n R(n, p_n) + n^2/(\log(n))^2 T(n, p_n))$, which is attractive in the scenario of $p_n \gg n$.

8.7 Problems

1. Prove Theorem 8.1.

2. Redo the simulation studies in Section 8.4.

3. Redo the simulation studies in Section 8.4 with the variable screening procedure HZ-SIS replaced by a sparse Bayesian neural network [96, 162].

4. Redo the simulation studies in Section 8.4 with the conditional independence test procedure replaced by the GAN-based test[3] [9].

5. Perform a simulation study with mixed data of Gaussian and binary, choosing appropriate variable screening procedures from the R package *MFSIS* [24] for use.

---

[3]The implementation of the test is available at `https://github.com/alexisbellot/GCIT`.

Chapter 9

High-Dimensional Inference with the Aid of Sparse Graphical Modeling

This chapter introduces the method of Markov neighborhood regression (MNR) [94, 160] for high-dimensional statistical inference. Here the inference can be interpreted in a general sense, which includes both variable selection and uncertainty quantification for the regression coefficients. With the aid of sparse graphical modeling for the covariates, the MNR method has successfully broken the high-dimensional statistical inference problem into a series of low-dimensional statistical inference problems, where the conventional statistical tests, such as t-test and Wald test, and their associated confidence interval construction methods can still be applied. The MNR method has also led to improved variable selection for high-dimensional linear and generalized linear models with the use of the dependency structure among the covariates. The MNR method has brought new insights into high-dimensional statistics, and it can be easily extended to nonlinear and non-Gaussian problems based on appropriate nonparametric conditional independence tests such as those discussed in Section 8.3.

9.1 Introduction

During the past two decades, the statistical literature has been overwhelmed by high-dimensional statistical research, including those on sparse graphical modeling, variable selection, and statistical inference (also known as "uncertainty quantification" in a narrow sense). Sparse graphical modeling has been discussed in the previous chapters of this book, and variable selection and uncertainty quantification will be the focus of this chapter.

The goal of variable selection is to identify the true explanatory variables for the response from a large set of suspicious ones under the sparsity assumption. A non-exhaustive list of the methods includes Lasso [164], SCAD [43], elastic net [206], group Lasso [192], adaptive Lasso [205], MCP [195], rLasso [149], Bayesian subset modeling [98], split-and-merge [150], and Bayesian shrinkage [151]. Toward uncertainty quantification for high-dimensional regression, a variety of methods have been developed with various degrees of success, including desparsified Lasso [71, 173, 196], multi sample-splitting [116], ridge projection [17], post-selection inference [11, 87], and residual-type bootstrapping [21, 104], among others.

DOI: 10.1201/9780429061189-9

A major issue with these methods is that they do not make much use of the dependency information among the covariates while conducting variable selection or uncertainty quantification for high-dimensional regression. As a result, the significance of each covariate is hard to be assessed and the true covariates are hard to be identified. Liang et al. [94] and Sun and Liang [160] address this issue by introducing the MNR method, where the significance of each covariate is assessed by conditioning on its Markov blanket. Moreover, under an appropriate sparsity assumption, the MNR method has successfully broken the high-dimensional inference problem into a series of low-dimensional inference problems where conventional statistical inference methods, such as t-test and Wald test and associated confidence interval construction methods, can still be applied. This use of the dependency information among covariates has led to tremendous success in variable selection and statistical inference for high-dimensional regression. It is interesting to point out that the MNR method can be viewed as a direct application of (equation 1.7) by letting $X_i = Y$ (the response variable of the regression); that is, the high-dimensional problem can be considered under the general paradigm of conditional independence tests.

In the remaining part of this chapter, we will first give a brief review of other high-dimensional inference methods and then describe the MNR method and illustrate it with some numerical examples.

9.2   Related Works

Consider the high-dimensional linear regression

$$Y = X\beta + \varepsilon, \tag{9.1}$$

where $Y \in \mathbb{R}^n$ and $X \in \mathbb{R}^{n \times p}$ represent the response vector and design matrix, respectively, $\beta \in \mathbb{R}^p$ represents the regression coefficients, and $\varepsilon \sim \mathcal{N}_n(0, \sigma^2 I_n)$ represents random errors. In this chapter, we generally assume that $n \ll p$ and the covariates X have been centered to zero, and we let $\hat{\Sigma} = X^T X / n$ denote the empirical estimator of the covariance matrix of the covariates.

9.2.1   Desparsified Lasso

The desparsified Lasso method was developed by Javanmard and Montanari [71], van de Geer et al. [173], and Zhang and Zhang [196], see also the review article [37]. Since the Lasso estimator of β is generally biased, a bias-corrected estimator is considered for the purpose of inference:

$$\hat{\beta}_{bc} = \hat{\beta}_{\text{Lasso}} + \hat{\Theta} X^T (y - X\hat{\beta}_{\text{Lasso}})/n, \tag{9.2}$$

where $\hat{\beta}_{\text{Lasso}}$ is the Lasso estimator, and $\hat{\Theta}$ is a concentration matrix (of X) representing an estimator of Σ^{-1}. From equation (9.2), one can derive that

$$\sqrt{n}(\hat{\beta}_{bc} - \beta) = \hat{\Theta} X^T \varepsilon / \sqrt{n} + \sqrt{n}(I_p - \hat{\Theta}\hat{\Sigma})(\hat{\beta}_{\text{Lasso}} - \beta)$$
$$= \hat{\Theta} X^T \varepsilon / \sqrt{n} + \Delta_n, \tag{9.3}$$

where I_p is an identity matrix of rank p, and $\Delta_n := \sqrt{n}(I_p - \hat{\Theta}\hat{\Sigma})(\hat{\beta}_{\text{Lasso}} - \beta)$ represents the remained bias in $\hat{\beta}_{bc}$. Therefore, if the concentration matrix $\hat{\Theta}$ is chosen appropriately such that $\|\Delta_n\|_\infty = o_p(1)$ holds, then $\sqrt{n}(\hat{\beta}_{bc} - \beta)$ asymptotically follows the Gaussian distribution $\mathcal{N}_p(0, \hat{\sigma}^2\hat{\Theta}\hat{\Sigma}\hat{\Theta}^T)$, where $\hat{\sigma}^2$ denotes a consistent estimate of σ^2. In practice, the concentration matrix $\hat{\Theta}$ can be estimated through node-wise regression [118]. Based on this result, confidence intervals can be constructed for components of β. However, as shown in Liang et al. [94] by numerical examples, these confidence intervals can be deficient in coverage rates when the condition $\|\Delta_n\|_\infty = o_p(1)$ is violated.

9.2.2 Ridge Projection

The ridge projection method [17] starts with the conventional ridge estimator:

$$\hat{\beta}_{\text{ridge}} = \frac{1}{n}(X^T X/n + \lambda I)^{-1} X^T Y,$$

where λ denotes the ridge regularization parameter. Let $P_R = X^T(XX^T)^- X$ be a projection matrix, where $(XX^T)^-$ denotes a generalized inverse of XX^T. To correct the bias of $\hat{\beta}_{\text{ridge}}$ [17], suggested the following estimator:

$$\hat{b}_R^{(j)} = \frac{\hat{\beta}_{\text{ridge}}^{(j)}}{P_R^{(j,j)}} - \sum_{k \neq j} \frac{P_R^{(j,k)}}{P_R^{(j,j)}} \hat{\beta}_{\text{Lasso}}^{(k)}, \quad j = 1, 2, \ldots, p,$$

where $\hat{\beta}_{\text{Lasso}}^{(k)}$ denotes the kth element of the Lasso estimator of β, and $P_R^{(j,k)}$ denotes the $(j,k)^{th}$ element of P_R. Under appropriate conditions, it can be shown that

$$\sigma^{-1}(\Omega_R^{(j,j)})^{-1/2}(\hat{b}_R^{(j)} - \beta^{(j)}) \approx (\Omega_R^{(j,j)})^{-1/2}\frac{W^{(j)}}{P_R^{(j,j)}} + \sigma^{-1}(\Omega_R^{(j,j)})^{-1/2}\Delta_R^{(j)}, \quad (9.4)$$

where $W := (W^{(1)}, \ldots, W^{(p)})^T \sim \mathcal{N}_p(0, \Omega_R)$, $\beta^{(j)}$ is the jth element of β, and

$$|\Delta_R^{(j)}| \leq \max_{k \neq j} \left| \frac{P_R^{(j,k)}}{P_R^{(j,j)}} \right| \left(\frac{\log(p)}{n} \right)^{1/2 - \xi},$$

with a typical choice $\xi = 0.05$. The confidence intervals for β can then be constructed based on equation (9.4). Bühlmann [17] also pointed out that the bias term $\Delta_R^{(j)}$ is typically not negligible and thus the Gaussian part of equation (9.4) needs to be corrected accordingly.

9.2.3 Post-Selection Inference

Instead of making inference for all regression coefficients, some authors proposed to make inference for the coefficients of the selected model only. That is, they aim to construct a valid confidence region $\widehat{CI}_{\hat{M}}$ such that

$$\lim_{n\to\infty} \inf P(\beta_{\hat{M}} \in \widehat{CI}_{\hat{M}}) \geq 1 - \alpha, \tag{9.5}$$

where $\hat{M}$ denotes a model or a set of covariates selected by a method, and α is the confidence level. However, $\hat{M}$ is random and depends on the data, which can make the problem rather complicated.

To address this issue, Lee et al. [87], Tibshirani et al. [165], and others (see Kuchibhotla et al. [83] for an overview and references therein) proposed to make conditional selective inference. They were to find the confidence region CI_M such that

$$P(\beta_M^{(j)} \in CI_M^{(j)} | \hat{M} = M) \geq 1 - \alpha,$$

where $\beta_M^{(j)}$ denotes the jth-component of β_M. This formulation of the problem avoids comparisons of the coefficients across different models $M \neq M'$, but it also makes the inference depend on the method used in model selection. See Lee et al. [87] for an inference procedure based on the Lasso selection method.

Alternative to conditional selective inference, Berk et al. [11], later extended by Bachoc et al. [5], proposed to make simultaneous inference for a set of covariates $Q = \{M : M \subseteq \{1, 2, \dots, p\}$ by constructing the set of confidence regions $\{\widehat{CI}_M : M \in Q\}$ such that

$$\lim_{n\to\infty} \inf P\left(\cap_{M\in Q}\{\beta_M \in \widehat{CI}_M\}\right) \geq 1 - \alpha,$$

which implies (equation 9.5) holds for any $\hat{M} \in Q$. The simultaneous inference requires the set Q to be specified independent of data and, moreover, $\widehat{CI}_M$ to be calculated for all $M \in Q$. Both requirements can limit the applications of the method. Refer to [83] for an overview.

The third method, but perhaps the oldest one, belonging to the class of post-selection inference methods is sample splitting, see [136] for its recent development. Unlike the other two methods, the sample splitting method is easy to implement: it is to first split the data into a training set and a test set and then select a model $\hat{M}$ based on the training set and make the inference for $\beta_{\hat{M}}$ based on the test set. Since $\hat{M}$ is independent of the test set, the conventional inference procedure and asymptotics developed for low-dimensional regression apply. The major issue with this method is that the inference is only valid conditioned on the model $\hat{M}$ that was selected on the training set [48], which might be inconsistent with the mechanism underlying the data generation process.

9.2.4   Other Methods

In the literature, there are also some other methods that are particularly developed for confidence interval construction and p-value evaluation for high-dimensional regression, such as residual-type bootstrapping [104, 21], group-bound [117], decorrelated score statistic [124], and recursive online-score estimation [146], among others.

The residual-type bootstrapping method suffers from the super-efficiency issue; the resulting confidence intervals of the zero regression coefficients tend to be too narrow. The group-bound method aims to construct confidence intervals for the l_1-norm $\|\beta_{c_k}\|$ of a group $c_k \subset \{1, 2, \ldots, p\}$ of covariates and often performs less well for a single regression coefficient. The decorrelated score statistic method works in a similar spirit to desparsified Lasso, but debiasing the C. R. Rao's score test statistic. The recursive online-score estimation method is an extension of the sample splitting method, where the samples are first split into a series of nonoverlapping chunks, and then variable selection is performed and score equations are constructed in a recursive way.

9.3 Markov Neighborhood Regression

Consider a generalized linear model (GLM) with the density function given by

$$f(y|x, \beta, \sigma) = h(y, \sigma) \exp\{(a(\theta)y - b(\theta))/c(\sigma)\}, \tag{9.6}$$

where σ is the dispersion parameter, θ is the canonical parameter that relates y to the covariates through a linear function

$$\theta = \beta_0 + X_1\beta_1 + \cdots + X_p\beta_p, \tag{9.7}$$

and the functions $a(\cdot)$, $b(\cdot)$, $c(\cdot)$ and $h(\cdot)$ are known. One can define different members of the family, such as Gaussian, binomial, Poisson and exponential, by specifying different functions $a(\cdot)$, $b(\cdot)$, $c(\cdot)$ and $h(\cdot)$.

By its conditional effect evaluation nature, statistical inference for a regression coefficient, say β_j, amounts to inference for the joint distribution $P(Y, X_j | X_{V\setminus\{j\}})$, where $V = \{1, 2, \ldots, p\}$ is the index set of the covariates included in equation (9.7). Let $G = (V, E)$ denote a sparse Markov graph formed by the covariates $\{X_1, X_2, \ldots, X_p\}$, where E denotes the edge set of the graph. Let ξ_j denote the Markov blanket of X_j, and let $\hat{\xi}_j$ denote a super Markov blanket of X_j such that $\xi_j \subseteq \hat{\xi}_j$. Let S^* denotes the index set of nonzero regression coefficients in equation (9.7), and let $\hat{S}^*$ denote a superset of S^* such that $S^* \subset \hat{S}^*$ holds. Then, similar to equation (1.7), Sun and Liang [160] shows

$$
\begin{aligned}
P(Y, X_j | X_{V\setminus\{j\}}) &= P(Y | X_j, X_{V\setminus\{j\}}) P(X_j | X_{V\setminus\{j\}}) \\
&= P(Y | X_{S^*}) P(X_j | X_{\xi_j}) \\
&= P(Y | X_{\hat{S}^*}) P(X_j | X_{\hat{\xi}_j}) \\
&= P(Y, X_j | X_{\hat{S}^* \cup \hat{\xi}_j}) \\
&= P(Y | X_{\hat{S}^* \cup \hat{\xi}_j \cup \{j\}}) P(X_j | X_{\hat{S}^* \cup \hat{\xi}_j}),
\end{aligned}
\tag{9.8}
$$

where $P(Y | X_{\hat{S}^* \cup \hat{\xi}_j \cup \{j\}})$ is modeled by a subset GLM with the canonical parameter given by

$$\theta = \beta_0 + X_j\beta_j + X_{\hat{S}^* \cup \hat{\xi}_j}^T \beta_{\hat{S}^* \cup \hat{\xi}_j}, \tag{9.9}$$

and $\beta_{\hat{S}^* \cup \hat{\xi}_j}$ denotes the regression coefficients of $X_{\hat{S}^* \cup \hat{\xi}_j}$.

Sun and Liang [160] pointed out that if a likelihood ratio test is used to test the hypothesis

$$H_0 : \beta_j = 0 \quad \text{versus} \quad H_1 : \beta_j \neq 0, \tag{9.10}$$

with respect to the GLM (equations 9.6 and 9.7), then, by equation (9.8), this test can be reduced to a likelihood ratio test with respect to the subset GLM (equations 9.6 and 9.9). Since the set $\hat{S}^* \cup \hat{\xi}_j$ forms a Markov neighborhood of X_j in the Markov graph G, this method is called Markov neighborhood regression (MNR) by Liang et al. [94] and Sun and Liang [160]. Further, suppose that the size of the Markov neighborhood satisfies the condition $|\hat{S}^* \cup \hat{\xi}_j \cup \{j\}| = o(n^{1/2})$ as n increases, then the test for the subset model can be conducted as for a low-dimensional problem with the conventional inference procedure and asymptotics held. The MNR method can be summarized in Algorithm 9.1. *With the aid of sparse graphical modeling for the covariates, the MNR method has broken the high-dimensional statistical inference problem for the GLM (equations 9.6 and 9.7) into a series of low-dimensional statistical inference problems.*

Algorithm 9.1 *(Markov Neighborhood Regression; [160])*

a. *(Variable selection) Select a subset of covariates for the model (9.6 and 9.7) such that $S^* \subseteq \hat{S}^*$ holds in probability as the sample size n diverges, where $\hat{S}^*$ indexes the selected covariates.*

b. *(Markov blanket estimation) Learn a Markov graph $\hat{G} = (V, \hat{E})$ such that $E \subseteq \hat{E}$ holds in probability as the sample size n diverges, where $\hat{E}$ denotes the edge set of the learnt Markov graph.*

c. *(Subset regression) For each covariate X_j, $j = 1, \ldots, p$, conduct a subset GLM $Y \sim X_{D_j}$, constructing the confidence interval for β_j and calculating the p-value for the hypothesis test (equation 9.10) as for a low-dimensional GLM.*

The validity of the MNR method can be established by the following lemma, whose proof follows directly from the above arguments and the asymptotics of the GLM with a diverging dimension [133].

Lemma 9.1 *Consider the high-dimensional GLM (equations 9.6 and 9.7). For any $j \in \{1, 2, \ldots, p\}$, let $D_j := \hat{S}^* \cup \hat{\xi}_j \cup \{j\}$, and assume the following conditions hold:*

$$S^* \subseteq \hat{S}^*, \tag{9.11}$$

$$\xi_j \subseteq \hat{\xi}_j, \tag{9.12}$$

$$|D_j| = o(\sqrt{n}). \tag{9.13}$$

Let x_{D_j} denote the design matrix of the subset model (9.6 and 9.9), let $\hat{\beta}_{D_j}$ be the MLE of their regression coefficients, and let $\hat{\beta}_j$ be the component of $\hat{\beta}_{D_j}$ corresponding to the covariate X_j. Then

$$\sqrt{n}(\hat{\beta}_j - \beta_j)/\sqrt{\hat{k}_{jj}} \sim N(0, 1), \tag{9.14}$$

where $\hat{k}_{jj}$ denotes the $(j,j)^{th}$ entry of the inverse of the observed information matrix $J_n(\hat{\beta}_{D_j}) = -\sum_{i=1}^{n} H_{\hat{\beta}_{D_j}}(\log f(y_i|\beta_{D_j}, x_{D_j}))/n$ and $H_{\hat{\beta}_{D_j}}(\cdot)$ denotes the Hessian matrix evaluated at $\hat{\beta}_{D_j}$.

Lemma 9.1 provides a theoretical basis for the MNR method. Condition (9.11) essentially requires the sure screening property to be held for the variable selection step of Algorithm 9.1. It is known that this property holds for many high-dimensional variable selection algorithms, such as Lasso [61, 164], SCAD [43], MCP [195], elastic net [206], and adaptive Lasso [205], under appropriate conditions.

Condition (9.12) essentially requires the sure screening property to be held for the Markov blanket estimation step. Therefore, step (b) of Algorithm 9.1 can be implemented by a few algorithms depending on the types of covariates. For Gaussian covariates, it can be implemented by the ψ-learning algorithm [93] as described in Chapter 2, nodewise regression [118], or graphical Lasso [49]. For mixed variables of Gaussian and binomial, it can be implemented by the p-learning method [183] as described in Chapter 6 or the nodewise regression method as described in [22].

Conditions (9.13) concern the sparsity of the GLM (equations 9.6 and 9.7) and the graphical model $G = (V, E)$. To cope with the sure screening property required by the variable selection and Markov blanket estimation steps, we can impose the following extra conditions on S^* and ξ_j's:

$$|S^*| \prec \min\{n/\log(p), \sqrt{n}\},$$
$$\max_{j \in \{1,2,...,p\}} |\xi_j| \prec \min\{n/\log(p), \sqrt{n}\}. \tag{9.15}$$

These conditions seem a little restrictive. A relaxation for them will be discussed in Section 9.3.2.

As a summary for the above discussion, we state the following theorem whose proof is provided in Sun and Liang [160].

Theorem 9.1 *(Validity of the MNR method; [160]) Consider the high-dimensional GLM (equations 9.6 and 9.7), whose covariates are mixed with Gaussian and binomial variables and interacted pairwisely. Suppose that the condition (9.15) is satisfied, a regularization method with an amenable penalty function is used for variable selection in step (a), and the nodewise regression method with an amenable penalty function is used for Markov blanket estimation in step (b). Then Algorithm 9.1 is valid for statistical inference of the high-dimensional GLM.*

Regarding this theorem, we have a few remarks. First, the amenable penalty function [105, 106] is general, which has included the Lasso penalty [164], SCAD penalty [43] and MCP penalty [195] as special cases. Second, a Gaussian version of the theorem, i.e., all X_j's are Gaussian, has been studied in Liang et al. [94]. Third, when Y is Gaussian, each subset model is reduced to a low-dimensional linear regression, for which the asymptotic normality approximation (9.14) can be replaced by the exact student t-distribution. Finally, when both Y and X_j's are Gaussian, Liang et al. [94] provided an interesting proof for the validity of the MNR method based on simple matrix operations.

9.3.1   Joint Statistical Inference

The MNR method described above can be easily extended to joint statistical inference for several regression coefficients. For example, to test the general linear
hypothesis

$$H_0 : C\beta = 0, \quad \text{versus} \quad H_1 : C\beta \neq 0, \tag{9.16}$$

where $C \in \mathbb{R}^{r \times p}$ is a matrix with a finite column rank of q and a finite row rank of
$r < n$, we can construct a subset GLM based on the joint Markov blankets of the
q covariates involved in the hypothesis. More precisely, let A_C denote the set of q
covariates involved in the hypothesis (equation 9.16) and let $D_C = A_C \cup (\cup_{j \in A_C} \hat{\xi}_j) \cup$
$\hat{S}^*$, then we can conduct the subset GLM $Y \sim X_{D_C}$ and test the hypothesis (equation
9.16) as for a low-dimensional GLM.

9.3.2   Non-Wald-Type Confidence Interval

As mentioned previously, the condition (9.13) is a little restrictive, which is required
for achieving the asymptotic normality (9.14). However, as pointed out by Sun and
Liang [160], this condition can be much relaxed if we do not stick to the Wald-type
confidence interval. For the case of normal linear regression, by Theorem 1 of He et
al. [64], we have

$$P\{-2\log L_n(C) > \chi_r^2(\alpha)\} \to \alpha \text{ if and only if } \lim_{n \to \infty} \frac{|D_C|}{n} = 0, \tag{9.17}$$

where $\log L_n(C)$ denotes the log-likelihood ratio corresponding to the hypothesis
(9.16), α is the test significance level, and $\chi_r^2(\alpha)$ is the upper α-quantile of a χ^2-
distribution with the degree of freedom r. This result implies that the condition (9.13)
can be relaxed to $|D_j| = o(n)$ by approximating the likelihood ratio test by a Chi-
square test. It is interesting to note that the approximation (9.17) also holds when r
increases with n, as long as $r < |D_C|$ and $\sqrt{r}(|D_C| - r/2) = o(n)$ hold.

For the case of logistic regression, by Sur et al. [163], the likelihood ratio test
for the hypothesis (9.16) can be approximated by a rescaled Chi-square test provided
that the covariates are Gaussian, $q \leq r$, and $\lim_{n \to \infty} |D_C|/n < \kappa$ for some constant
$0 < \kappa < 1/2$.

Under these scenarios, the condition (9.15) can be much relaxed as

$$|S^*| \prec \frac{n}{\log(p)}, \quad \max_{j \in \{1,2,\dots,p\}} |\xi_j| \prec \frac{n}{\log(p)}.$$

Further, if one assumes that the dimension p increases with the sample size n at
a polynomial rate, i.e., $p = O(n^a)$ for some constant $a > 0$, then the relaxed condition implies that the variable selection procedure, performed in steps (a) and (b)
of Algorithm 9.1, can be replaced by a sure independence screening (SIS) procedure [45, 46]. This replacement might significantly accelerate the computation of the
MNR method.

9.4 An Illustrative Example

We first illustrate the concept of MNR by comparing it with the ordinary least square (OLS) method on a low-dimensional problem. A similar example has been considered in Liang et al. [94]. We simulated a dataset from the normal linear regression (equation 9.1), where we set the sample size $n = 5,000$, the dimension $p = 100$, the true value of $\sigma^2 = 1$, the true regression coefficients $(\beta_0, \beta_1, \beta_2, \ldots, \beta_5) = (1, 0.3, 0.5, -0.5, -0.3, 0.8)$, and $\beta_6 = \cdots = \beta_p = 0$. The covariates were simulated from a multivariate Gaussian distribution $\mathcal{N}_p(0, \Sigma)$ with a Toeplitz covariance matrix

$$\Sigma_{i,j} = 0.9^{|i-j|}, \quad i, j = 1, \ldots, p.$$

The MNR method was implemented with the R package *SIS* [141], where we used SIS-SCAD for variable selection and nodewise regression (with SIS-Lasso for each node) for Markov blanket estimation. The SIS-SCAD refers to that one first screens the variables by a SIS procedure and then selects variables from the survived ones with the SCAD penalty. SIS-Lasso and SIS-MCP (used later) can be interpreted in the same way. Figure 9.1 compares 95% confidence intervals resulted from MNR and OLS for a simulated linear regression problem: the two methods produced almost identical confidence intervals for each regression coefficient. This experiment confirms the validity of the MNR method.

9.5 Simulation Studies: Confidence Interval Construction

This example was modified from one example of Liang et al. [94]. We considered three types of GLM models, linear regression, logistic regression and Cox regression. For each type of model, we generated covariates from the same zero-mean multivariate Gaussian distribution with the concentration matrix as given in (2.10), while varying the sample size and regression coefficients.

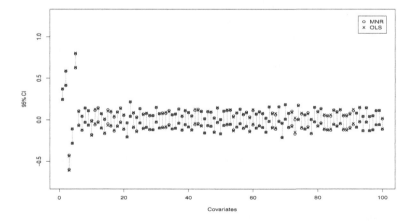

Figure 9.1 *Comparison of MNR and OLS for their 95% confidence intervals constructed for a low-dimensional linear regression problem with* $n = 5000$ *and* $p = 100$.

9.5.1   Linear Regression

We simulated 100 datasets independently from the linear regression (9.1), where we set $n = 200$, $p = 500$, $\sigma^2 = 1$, and the regression coefficients $(\beta_0, \beta_1, \beta_2, \ldots, \beta_5, \beta_6, \ldots, \beta_p) = (1, 2, 2.25, 3, 3.25, 3.75, 0, \ldots, 0)$. For the MNR method, we used SIS-MCP for variable selection and used the ψ-learning method for Markov blanket estimation. For ψ-learning, we set $\alpha_1 = 0.1$ and $\alpha_2 = 0.05$ for its correlation screening and ϕ-screening steps, respectively. The same settings of α_1 and α_2 were also applied to the logistic and Cox regression datasets below.

9.5.2   Logistic Regression

We simulated 100 datasets independently from the logistic regression, where we set $n = 300$, $p = 500$, $(\beta_0, \beta_1, \ldots, \beta_5, \beta_6, \ldots, \beta_p) = (1, 2, 2.25, 3, 3.25, 3.75, 0, \ldots, 0)$, and set the ratio of the cases and control samples to be 1:1. As for the linear regression case, we employed SIS-MCP for variable selection and the ψ-learning method for Markov blanket estimation.

9.5.3   Cox Regression

We simulated 100 datasets independently from the Cox regression

$$\lambda(t) = \lambda_0 \exp(\beta_1 X_1 + \beta_2 X_2 + \ldots + \beta_p X_p), \tag{9.18}$$

where we set $n = 400$, $p = 500$, $(\beta_1, \ldots, \beta_5, \beta_6, \ldots, \beta_p) = (1, 1, 1, 1, 1, 0, \ldots, 0)$, the baseline hazard rate $\lambda_0 = 0.1$, and the censoring hazard rate $\lambda_c = 1$. For the MNR method, we used SIS-Lasso for variable selection and the ψ-learning method for Markov blanket estimation.

Table 9.1 summarizes the numerical results produced by MNR, desparsified Lasso, and ridge projection for the simulated data. The implementations of desparsified Lasso and ridge projection are available in the R package *hdi* [114]. The comparison indicates the superiority of MNR, which does not only produce accurate coverage rates for linear regression and logistic regression but also works well for Cox regression. The desparsified Lasso and ridge projection methods basically fail for the examples.

9.6   Simulation Studies: Variable Selection

As a part of statistical inference, MNR can also be used for variable selection. For this purpose, we can simply conduct a multiple hypothesis test for the hypotheses:

$$H_{0,i} : \beta_i = 0, \quad H_{1,i} : \beta_i \neq 0, \quad i = 1, 2, \ldots, p, \tag{9.19}$$

based on the p-values evaluated in step (c) of Algorithm 9.1. The multiple hypothesis test will cluster the covariates into two groups, which correspond to the null and alternative hypotheses in equation (9.19), respectively.

Table 9.1 *Comparison of 95% confidence intervals resulted from MNR, desparsified Lasso and ridge projection for the simulated examples.*

Model	Measure		Desparsified Lasso	Ridge	MNR
Linear	Coverage	Signal	0.2600	0.3500	0.9580
		Noise	0.9635	0.9919	0.9496
	Width	Signal	0.2757 (0.0031)	0.4437 (0.0044)	0.2784 (0.0022)
		Noise	0.2658 (0.0027)	0.4274 (0.0037)	0.2781 (0.0023)
Logistic	Coverage	Signal	0.0160	0	0.9480
		Noise	0.9946	0.9999	0.9393
	Width	Signal	0.6337 (0.01196)	1.0621 (0.0113)	1.809 (0.0446)
		Noise	0.5730 (0.0100)	0.9958 (0.0097)	0.9518 (0.0109)
Cox	Coverage	Signal	—	—	0.9100
		Noise	—	—	0.9365
	Width	Signal	— —	— —	0.2905 (0.0014)
		Noise	— —	— —	0.2309 (0.0012)

We suggest to conduct the multiple hypothesis test using the empirical Bayesian method by Liang and Zhang [95]. To facilitate the test, we first convert the p-values to z-scores through the transformation:

$$z_i = \Phi^{-1}(1 - p_i), \tag{9.20}$$

where p_i denotes the p-value of covariate i, and $\Phi(\cdot)$ denotes the CDF of the standard Gaussian distribution. Then we fit the z-scores by a mixture Gaussian or, more generally, a mixture exponential power distribution, which separates the z-scores to two groups, significantly or non-significantly different from zero. The two groups of z-scores correspond to the two groups of covariates with zero or nonzero regression coefficients, respectively.

Figure 9.2 shows the histograms of the z-scores calculated with the datasets simulated in Section 9.5, which indicate that the true and false covariates are well separated in z-scores. Table 9.2 compares MNR with SIS-SCAD, SIS-MCP and SIS-Lasso in variable selection on those datasets. The comparison indicates that MNR can provide a drastic improvement over the other methods in variable selection. In particular, with an appropriate choice of the q-value, say, 0.001, the MNR method performs almost perfectly in variable selection for the simulated datasets. In contrast, since the penalty used in SIS-SCAD, SIS-MCP and SIS-Lasso generally shrinks the regression coefficients toward zero, these regularization methods tend to select more

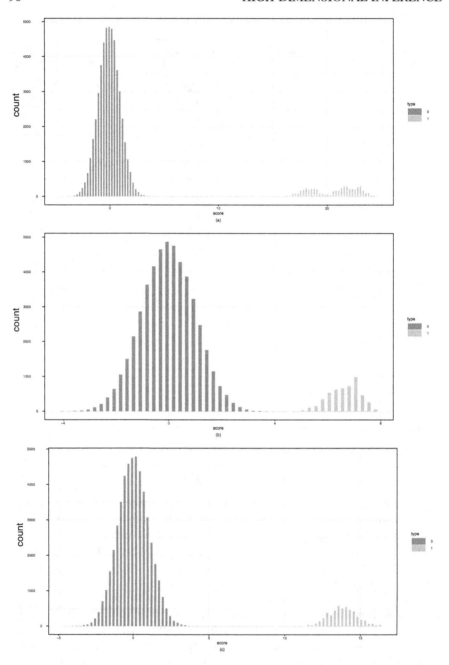

Figure 9.2 *Histograms of z-scores generated by the MNR method for the simulated example, where the panel (a), (b), (c) plots are for the linear, logistic, and Cox regression, respectively, and the z-scores of the true covariates have been duplicated ten times for a balanced view of the plots.*

Table 9.2 *Variable selection for the linear, logistic and cox regression datasets simulated in Section 9.5, where q denotes the significance level (measured in Storey's q-value [156]) of the multiple hypothesis test.*

Model	Measure	MNR(q) 0.0001	0.001	0.01	SIS-SCAD	SIS-MCP	SIS-Lasso
Linear	FSR	0.0013	0.0027	0.0145	0.0040	0.0654	0.8654
	NSR	0	0	0	0	0	0.004
Logistic	FSR	0	0	0.0177	0.5187	0.3353	0.6033
	NSR	0.006	0	0	0.2260	0.1080	0.2860
Cox	FSR	0	0.0040	0.0272	—	—	0.6919
	NSR	0	0	0	—	—	0.014

false variables for compensation of the shrinkage effect. As shown in Table 9.2, SIS-SCAD, SIS-MCP and SIS-Lasso tend to have a high false selection rate (FSR). See Section 9.9 for the definitions of FSR and negative selection rate (NSR).

9.7   Causal Structure Discovery

The causal relationship is usually referred to as cause and effect between two variables in a dataset, which exists when one variable has a direct influence on the other. In other words, a pair of variables will be considered to have no causal relationship if they are conditionally independent given other variables in the dataset.

Amounting to a series of conditional independence tests, the PC algorithm [153] was developed for learning the causal structure of a Bayesian network, and later it was extended by Bühlmann et al. [18] for identifying causal covariates for high-dimensional linear regression by assuming that the response variable is the effect variable of covariates. Under the same assumption, Liang et al. [94], Liang and Liang [101], and Sun and Liang [160] extended the MNR method to identify the causal covariates for high-dimensional linear, GLM and nonlinear regressions, respectively. Since, in this case, we are not interested in making inference for the effect of each covariate, the computation of the MNR method can be simplified by restricting the inference to the covariates that survive the sure screening process. In summary, we can have the following algorithm for causal structure discovery:

Algorithm 9.2 *(Simplified MNR for Causal Structure Discovery; [94])*

a. *(Variable screening) Apply a SIS procedure to the regression $Y \sim X$ to obtain a reduced covariate set, $\tilde{S}^* \subseteq \{1,\ldots,p\}$, with the size $|\tilde{S}^*| = O(n/\log(n))$.*

b. *(Markov neighborhood estimation) For each covariate $X_j \in \tilde{S}_*$, apply a SIS procedure to the regression $X_j \sim X_{V \setminus \{j\}}$ to obtain a reduced neighborhood $\tilde{\xi}_j \subseteq \{1,\ldots,p\}$ with the size $|\tilde{\xi}_j| = O(n/\log(n))$.*

c. *(Subset Regression) For each covariate $X_j \in \tilde{S}^*$, run a subset regression with the covariates given by $X_{\{j\} \cup \tilde{S}^* \cup \tilde{\xi}_j}$, calculating the p-value for the hypothesis test $H_0 : \beta_j = 0$ versus $H_1 : \beta_j \neq 0$.*

Table 9.3 *Drug-sensitive genes were identified by MNR for four selected drugs, where the number in the parentheses denotes the width of the 95% confidence interval of the corresponding regression coefficient.*

Chemical compound	Drug-sensitive genes
17-AAG	NQO1 (0.115)
Irinotecan	ARHGAP19 (0.108), SLFN11 (0.033)
Paclitaxel	BCL2L1 (0.289)
Topotecan	SLFN11 (0.107)

The contents of the table were taken from Table 9 of Liang et al. [94].

d. *(Causal Structure Discovery) Identify the causal covariates among $\tilde{S}^*$ by conducting a multiple hypothesis test based on the p-values calculated in step (c).*

The consistency of the algorithm can be justified by following the discussion in Section 9.3.2. To illustrate Algorithm 9.2, Liang et al. [94] considered the problem of drug-sensitive gene identification with the cancer cell line encyclopedia (CCLE) dataset[1]. Their results are briefly reviewed as follows.

The dataset 24 chemical compounds. For each chemical compound, the dose response was measured at 8 points, and the activity areas [8], i.e., the areas under the dose-response curves, are used as the response variable of the regression, while the covariates consist of expression values of $p = 18,988$ genes. For each chemical compound, the number of cell lines, i.e., the sample size n, varies slightly around 400. Liang et al. [94] applied Algorithm 9.2 to this dataset, where the Henze-Zirkler sure independence screening (HZ-SIS) algorithm [184] was used for variable screening and Markov neighborhood estimation. For both steps, the neighborhood size was set to 40, which makes the student t-test feasible for calculating the p-values in step (c). In step (d), the multiple hypothesis test was done with the adjusted p-values [66] at a significance level of 0.05; that is, the genes with the adjusted p-values ≤ 0.05 were identified as drug-sensitive genes. Part of their results is shown in Table 9.3. Refer to [94] for more results.

For comparison, Liang et al. [94] also applied the desparsified Lasso and ridge projection method to this example. Unfortunately, both methods did not work well for the problem because of the high dimensionality of the dataset. For desparsified Lasso, the package *hdi* aborted to the excess of memory limit; and for ridge projection, *hdi* also aborted for some drugs.

Liang et al. [94] noted that many of the drug-sensitive genes identified by the MNR method have been verified in the literature. For example, MNR identified the gene SLFN11 for both drugs Topotecan and Irinotecan, which has been verified in Barretina et al. [8] and Zoppoli et al. [204] that SLFN11 is predictive of treatment response for the two drugs. For the drug 17-AAG, MNR identified the gene NQO1, which has been verified in Barretina et al. [8] and Hadley and Hendricks [58] that NQO1 is the top predictive biomarker for 17-AAG. For the drug Paclitaxel, MNR

---

[1]The dataset is publicly available at www.broadinstitute.org/ccle.

identified the gene BCL2L1; it has been reported in the literature, see e.g., [40, 86], that BCL2L1 is predictive of treatment response for Paclitaxel.

9.8 Computational Complexity

The MNR method has a computational complexity of $O(n^3 p^2)$, since it needs to estimate the Markov graph for p covariates. For example, if the nodewise Lasso regression is used for estimating the Markov network, a total of p regressions need to be performed, and each has a computational complexity of $O(n^3 p)$ [115]. As analyzed in Chapter 2, the ψ-learning method has also about the same computational complexity in estimating the Markov network. However, this computational complexity should not be a big concern for the efficiency of the method, since MNR has an embarrassingly parallel structure which makes its computation easily accelerated with a parallel architecture.

9.9 Problems

1. Suppose that $(X_1, X_2, \ldots, X_p)^T \sim \mathcal{N}_p(\mathbf{0}, \Theta_p^{-1})$ and the first d variables $(X_1, X_2, \ldots, X_d)^T$ follow the marginal distribution $\mathcal{N}_d(\mathbf{0}, \Theta_d^{-1})$. Prove that Θ_p and Θ_d have the same $(1,1)^{th}$ entry if $\{X_2, X_3, \ldots, X_d\}$ forms a Markov blanket of X_1.

2. Prove Lemma 9.1.

3. Prove Theorem 9.1.

4. Redo the simulation example in Section 9.4.

5. Redo the confidence interval construction examples in Section 9.5.

6. Redo the variable selection examples in Section 9.6.

7. Redo the confidence interval construction examples in Section 9.5 with the nodewise MCP algorithm used for Markov blanket estimation.

8. Redo the variable selection examples in Section 9.6 with the nodewise MCP algorithm used for Markov blanket estimation.

9. Perform a simulation study with the covariates being mixed by Gaussian and binary variables, constructing confidence intervals, assessing p-values and selecting an appropriate subset model for the regression.

Appendix

This appendix gives some distributions, formulas, definitions, and codes that are used in previous chapters of the book.

A.1 Multivariate Gaussian Distribution

A random vector $X \in \mathbb{R}^p$ is said to follow a multivariate Gaussian distribution $\mathcal{N}_p(\mu, \Sigma)$ if it has the density function given by

$$f(x) = \frac{1}{(2\pi)^{p/2}|\Sigma|^{1/2}} e^{-(x-\mu)^T \Sigma^{-1}(x-\mu)/2},$$

where μ and Σ are called the mean vector and covariance matrix of the distribution, respectively.

If X_a and X_b are jointly multivariate Gaussian such that

$$\begin{pmatrix} X_a \\ X_b \end{pmatrix} \sim \mathcal{N}_p \left(\begin{pmatrix} \mu_a \\ \mu_b \end{pmatrix}, \begin{pmatrix} \Sigma_{aa} & \Sigma_{ab} \\ \Sigma_{ba} & \Sigma_{bb} \end{pmatrix} \right), \tag{A.1}$$

then the conditional distribution of X_b given X_a is multivariate Gaussian with the mean vector and covariance matrix given by

$$\begin{aligned} E(X_b|X_a) &= \mu_b + \Sigma_{ba}\Sigma_{aa}^{-1}(X_a - \mu_a), \\ \mathrm{Cov}(X_b|X_a) &= \Sigma_{bb} - \Sigma_{ba}\Sigma_{aa}^{-1}\Sigma_{ab}. \end{aligned} \tag{A.2}$$

A.2 The Basics of Graphical Models

The graphical model can be represented by an undirected graph $G = (V, E)$, where $V = \{1, 2, \ldots, p\}$ denotes the set of nodes with each representing a random variable, and $E = (e_{ij})$ denotes the adjacency matrix of the graph with each element $e_{ij} \in \{0, 1\}$ indicating the edge status between two nodes.

- *Adjacent*: Two nodes X_i and X_j in a graph are said to be adjacent if $e_{ij} = 1$.
- *Neighboring set*: it is defined as the set $n_G(i) = \{k : e_{ik} = 1\}$ for node X_i. The neighboring set is also known as neighborhood or boundary set.
- *Path*: A path from node X_i to node X_j is a sequence of distinct nodes connected by edges.
- *Separator*: For graphical models, a subset U is said to be a separator of the subsets I and J if for any node X_i in I and any node X_j in J, all paths from X_i to X_j pass through U.

A.3 Power-Law Distribution

A non-negative random variable U is said to follow a power-law distribution [81] if

$$P(U = u) \propto U^{-\lambda},$$

for some $\lambda > 0$. The connectivity of a network following a power-law distribution means that the majority of its nodes are of very low degree, while some nodes are of much higher degree.

A.4 Precision-Recall Curve

The precision-recall (PR) curve is a plot of Precision versus Recall defined by

$$\text{Precision} = \frac{\text{TP}}{\text{TP} + \text{FP}}, \quad \text{Recall} = \frac{\text{TP}}{\text{TP} + \text{FN}},$$

where TP, FP, and FN are as defined in Table A.1 for an experiment with P positive cases and N negative cases. The PR curve and the area under the curve (AUC) are often used in assessing the performance of a binary decision algorithm.

A.5 Receiver Operating Characteristic Curve

The receiver operating characteristic (ROC) curve is a plot of False Positive Rate (FPR) versus True Positive Rate (TPR) defined by

$$\text{FPR} = \frac{\text{FP}}{\text{FP} + \text{TN}}, \quad \text{TPR} = \frac{\text{TP}}{\text{TP} + \text{FN}},$$

where TP, FP, and FN are as defined in Table A.1.

Like the PR curve, the ROC curve and the AUC are often used in assessing the performance of a binary decision algorithm.

A.6 False and Negative Selection Rates

The false selection rate (FSR) and negative selection rate (NSR) are usually used to measure the accuracy of feature selection for high-dimensional regression. Let S^* denote the set of true features of a high-dimensional regression, and let $\hat{S}$ denote the set of features selected by a method. The FSR and NSR are defined by

$$\text{FSR} = \frac{|\hat{S} \setminus S^*|}{|\hat{S}|}, \quad \text{NSR} = \frac{|S^* \setminus \hat{S}|}{|S^*|},$$

Table A.1 *Possible outcomes for a binary decision problem.*

Decision \ Truth	Positive	Negative
Positive	True positive (TP)	False positive (FP)
Negative	False negative (FN)	True negative (TN)

where $|A|$ denotes the cardinality of the set A.

In simulation studies, one often has multiple estimates of S^* obtained from different runs of a method, where each run might work with a different dataset. In this scenario, FSR and NSR are defined as

$$\text{FSR} = \frac{\sum_{j=1}^{M} |\hat{S}_j \setminus S^*|}{\sum_{j=1}^{M} |\hat{S}_j|}, \quad \text{NSR} = \frac{\sum_{j=1}^{M} |S^* \setminus \hat{S}_j|}{M|S^*|},$$

where $\hat{S}_1, \hat{S}_2, \ldots, \hat{S}_M$ denote the multiple estimates of S^*.

A.7 A Multiple Hypothesis Testing Procedure

Consider a multiple hypothesis test for a total of N hypotheses:

$$H_{0,i} : \theta_i \in \Theta_0, \quad \text{versus} \quad H_{1,i} : \theta_i \notin \Theta_0, \quad i = 1, 2, \ldots, N. \tag{A.3}$$

Let $p_1, p_2, \ldots, p_N$ denote individual p-values of the N tests, and let

$$z_i = \Phi^{-1}(1 - p_i), \quad i = 1, 2, \ldots, N,$$

denote the z-scores, where $\Phi(\cdot)$ denotes the cumulative distribution function (CDF) of the standard normal distribution.

To conduct the multiple hypothesis test, Liang and Zhang [95] proposed the following empirical Bayesian procedure. It models the z-scores by an m-component mixture exponential power distribution:

$$g(z; \vartheta) = \sum_{i=1}^{m} \varpi_i f(z; v_i, \sigma_i, \alpha_i), \tag{A.4}$$

where $\vartheta = (\varpi_1, \varpi_2, \ldots, \varpi_{m-1}; v_1, \sigma_1, \alpha_1, v_2, \sigma_2, \alpha_2, \ldots, v_m, \sigma_m, \alpha_m)^T$ is the vector of parameters of the distribution. The parameters ϖ_i's are mixture proportions, which satisfy the constraints $0 < \varpi_i < 1$ and $\sum_{i=1}^{m} \varpi_i = 1$. The density function of the exponential power distribution is given by

$$f(z; v_i, \sigma_i, \alpha_i) = \frac{\alpha_i}{2\sigma_i \Gamma(1/\alpha_i)} e^{-(|z-v_i|/\sigma_i)^{\alpha_i}}, \quad -\infty < v_i < \infty, \ \sigma_i > 0, \ \alpha_i > 1, \tag{A.5}$$

where v_i, σ_i and α_i represent the center, dispersion and decay rate of the distribution, respectively. If $\alpha_i = 2$, the distribution (equation A.5) is reduced to the Gaussian distribution $N(v_i, \sigma_i^2/2)$, while $1 < \alpha_i < 2$ and $\alpha_i > 2$ correspond to the heavy-tailed and light-tailed cases, respectively. Refer to Holzmann et al. [67] for the proof of identifiability of the mixture distribution (equation A.4).

One can expect that the null and alternative effects contained in the hypotheses (equation A.3) can be modeled by the mixture distribution on the z-scores, where the most-left component (in mean value) corresponds to the null effect and the other components correspond to the alternative effect. We often set $m = 2$ or 3. Refer to [95] for more discussions on the choice of m.

The parameters of the mixture model can be estimated by minimizing the Kullback-Leibler divergence

$$\mathrm{KL}(g_\vartheta, \tilde{g}) = -\int \log\left\{\frac{g(z|\vartheta)}{\tilde{g}(z)}\right\} \tilde{g}(z)dz,$$

using the stochastic gradient descent (SGD) algorithm, where $\tilde{g}(z)$ denotes the true density function of z_i's. Refer to [95] for detail. This procedure of parameter estimation permits general dependence between test scores.

The cutoff value z_c for the multiple hypothesis test (equation A.3) can be determined by controlling the q-value [156] of the test. For a given clustering rule $\Lambda_c = \{z_i \geq z_c\}$, the q-value is defined by

$$q(z_c) := \inf_{\{\Lambda_c : z \in \Lambda_c\}} \mathrm{FDR}(\Lambda_c),$$

$$\mathrm{FDR}(\Lambda_c) := \frac{N\hat{\omega}_1[1 - F(z_c; \hat{v}_1, \hat{\sigma}_1, \hat{\alpha}_1)]}{\#\{z_i : z_i \geq z_c\}},$$

where $F(\cdot)$ is the CDF of the exponential power distribution (equation A.5), and $\mathrm{FDR}(\Lambda_c)$ denotes an estimate for the false discovery rate (FDR) among the cases in the set $\Lambda_c = \{z_i \geq z_c\}$. For example, if one wants to set the significance level of the multiple hypothesis test (equation A.3) to 0.05, then one can choose the cutoff value z_c such that $q(z_c) \leq 0.05$.

A.8 Codes for Some Simulation Studies

This section gives the codes used for producing some numerical results presented in the book.

A.8.1 Section 2.4

The code for producing the results of AR(2) in Table 2.1:

```
library (equSA)
library (huge)
library (PRROC)
library (flux)

n<-200; p<-300
Area1<-NULL; Area2<-NULL; Area3<-NULL

for (M in 1:10){
  #### Data Generation #####
  x<-GauSim(n=200,p=300,graph="AR(2)")

  #### \psi-learning #####
  edge1<-NULL; edge0<-NULL; m<-0
```

```
   for(i in 1:(p-1)){
     for(j in (i+1):p){
      m<-m+1
      if(x$theta[i,j]==1) edge1<-rbind(edge1, c(i,j,m))
         else edge0<-rbind(edge0,c(i,j,m))
       }
     }

  g<-equSAR(x$data, ALPHA1=0.05, ALPHA2=0.05)
  s<-g$score
  s1<-(-1)*s[edge1[,3],3]
  s0<-(-1)*s[edge0[,3],3]
  pr<-pr.curve(s1,s0,curve=TRUE)
  Area1<-c(Area1,pr$auc.integral)

#### GLasso #####
  h1<-huge(x$data,nlambda=100, method="glasso")
  R<-NULL
  for(k in 1:100){
    E1<-as.numeric(h1$path[[k]])
    B<-x$theta+E1
    L<-length(which(B==2))

    precision<-1.0*L/sum(E1)
    recall <-1.0*L/sum(x$theta)
    R<-rbind(R, c(recall,precision))
   }
  R<-rbind(R,c(0,1)); R<-rbind(R,c(1,0))
  auc.g<-auc(R[,1],R[,2])
  Area2<-c(Area2,auc.g)

  #### nodewise regression #####
  h2<-huge(x$data,nlambda=100, method="mb")
  R<-NULL
  for(k in 1:100){
    E2<-as.numeric(h2$path[[k]])
    B<-x$theta+E2
    L<-length(which(B==2))

    precision<-1.0*L/sum(E2)
    recall <-1.0*L/sum(x$theta)
    R<-rbind(R, c(recall,precision))
   }
  R<-rbind(R,c(0,1)); R<-rbind(R,c(1,0))
  auc.mb<-auc(R[,1],R[,2])
```

```
  Area3 <-c ( Area3 , auc .mb)
  }

Area<-cbind ( Area1 , Area2 , Area3 )
apply ( Area , 2 , mean )
apply ( Area , 2 , sd )/ sqrt (10)
t . test ( Area [ ,1] , Area [ ,2] , paired=T, alternative =" greater ")
t . test ( Area [ ,1] , Area [ ,3] , paired=T, alternative =" greater ")
```

A.8.2   Section 3.4

The code for producing the results with $p = 100$ in Table 3.1:

```
library (equSA)
library (spaceExt )
library (PRROC)
library ( flux )

samsize <-200;  dim<-100
Area1 <-NULL;   Area2 <-NULL

for (M in  1:10){
  #### Data  generation  #####
  x<-SimGraDat ( n=samsize , p=dim,  type=" band " ,  rate =0.1)

  #### \psi-learning  #####
  Est <-GraphIRO( x$data ,  x$A,  iteration =30,  warm=10)
  R<-Est$RecPre ;  R<-rbind (R, c (0 ,1)); R<-rbind (R, c (1 ,0))
  auc . IRO<-auc (R [ ,1] ,  R [ ,2])
  Area1 <-c ( Area1 , auc . IRO)

  #### missGlasso  #####
  R<-NULL
  lam<-seq (1.0 e-3 ,0.9 , len =50)
  for(s in  lam){
     g<-glasso . miss ( x$data , emIter =25 , rho=s )
     adjacency  <- abs(g$wi) > 1E-4;  diag ( adjacency ) <- 0
     B<-x$A+adjacency
     L<-length ( which (B==2))
     precision <-1.0*L/ sum( adjacency )
     recall <-1.0*L/ sum(x$A)
     R<-rbind (R,  c ( recall , precision , s , g$bic ))
  }
  R<-rbind (R, c (0 ,1)); R<-rbind (R, c (1 ,0))
  auc . mg<-auc (R [ ,1] ,R [ ,2])
```

```
  Area2 <-c ( Area2 , auc .mg)
  }
Area <-cbind ( Area1 , Area2 )
apply ( Area ,2 , mean )
apply ( Area ,2 , sd )/ sqrt (10)
```

A.8.3 Section 4.3

The code for producing the results with $p = 100$ in Table 4.1:

```
library ( equSA )
library (MASS)
library ( flux )
library (PRROC)
library ( huge )


dim <-100
c1 < -0.6;  c2 < -0.5;  c3 <-0.4
mu1 < -0.5;  mu2 <-0;  mu3 <--0.5

Sigma1 <-matrix ( rep (0 , dim*dim ),  ncol=dim )
for ( i  in  1:dim )  Sigma1 [ i , i ]<-1
for ( i  in  1:( dim -1)){  Sigma1 [ i , i+1]<-c1 ;  Sigma1 [ i+1,i ]
    <-c1 }
for ( i  in  1:( dim -2)){  Sigma1 [ i , i+2]<-c1 /2 ;  Sigma1 [ i+2,i ]
    <-c1 /2 }


Sigma2 <-matrix ( rep (0 , dim*dim ),  ncol=dim )
for ( i  in  1:dim )  Sigma2 [ i , i ]<-1
for ( i  in  1:( dim -1)){  Sigma2 [ i , i+1]<-c2 ;  Sigma2 [ i+1,i ]
    <-c2 }
for ( i  in  1:( dim -2)){  Sigma2 [ i , i+2]<-c2 /2 ;  Sigma2 [ i+2,i ]
    <-c2 /2 }


Sigma3 <-matrix ( rep (0 , dim*dim ),  ncol=dim )
for ( i  in  1:dim )  Sigma3 [ i , i ]<-1
for ( i  in  1:( dim -1)){  Sigma3 [ i , i+1]<-c3 ;  Sigma3 [ i+1,i ]
    <-c3 }
for ( i  in  1:( dim -2)){  Sigma3 [ i , i+2]<-c3 /2 ;  Sigma3 [ i+2,i ]
    <-c3 /2 }


A <-matrix ( rep (0 , dim*dim ), ncol=dim )
for ( i  in  1:( dim -1)){  A[ i , i+1]<-1; A[ i+1,i ]<-1 }
for ( i  in  1:( dim -2)){  A[ i , i+2]<-1; A[ i+2,i ]<-1 }
```

```
edge1<-NULL; edge0<-NULL; m<-0
for(i in 1:(dim-1)){
  for(j in (i+1):dim){
    m<-m+1
    if(A[i,j]==1) edge1<-rbind(edge1, c(i,j,m))
      else edge0<-rbind(edge0,c(i,j,m))
  }
}


Area<-NULL
for(M in 1:10){

  area <-NULL
  #### Data generation #####
  x1<-mvrnorm(n=100, rep(mu1,dim), solve(Sigma1))
  x2<-mvrnorm(n=100, rep(mu2,dim), solve(Sigma2))
  x3<-mvrnorm(n=100, rep(mu3,dim), solve(Sigma3))
  X<-rbind(x1,x2,x3)

  #### mixture \psi-learng #####
  g<-GGMM(X,A,3)
  R<-g$RecPre; R<-rbind(R,c(0,1)); R<-rbind(R,c(1,0))
  auc.m<-auc(R[,1], R[,2])
  area <-c(area,auc.m)

  #### \psi-learning #####
  g<-equSAR(X)
  s<-g$score
  s1 <-(-1)*s[edge1[,3],3]
  s0 <-(-1)*s[edge0[,3],3]
  pr<-pr.curve(s1,s0,curve=TRUE)
  area<-c(area,pr$auc.integral)

  #### GLasso ###########
  h1<-huge(X,nlambda=100, method="glasso")
  R<-NULL
  for(k in 1:100){
    E1<-as.numeric(h1$path[[k]])
    B<-A+E1
    L<-length(which(B==2))
    precision <-1.0*L/sum(E1)
    recall <-1.0*L/sum(A)
    R<-rbind(R, c(recall,precision))
  }
```

```
   R<-rbind(R,c(0,1)); R<-rbind(R,c(1,0))
   auc.g<-auc(R[,1],R[,2])
   area <-c(area,auc.g)

   #### nodewise regression #####
   h2<-huge(X,nlambda=100, method="mb")
   R<-NULL
   for(k in 1:100){
      E2<-as.numeric(h2$path[[k]])
      B<-A+E2
      L<-length(which(B==2))
      precision <-1.0*L/sum(E2)
      recall <-1.0*L/sum(A)
      R<-rbind(R, c(recall,precision))
   }
   R<-rbind(R,c(0,1)); R<-rbind(R,c(1,0))
   auc.mb<-auc(R[,1],R[,2])
   area <-c(area,auc.mb)

   Area<-rbind(Area,area)
}
apply(Area,2,mean)
apply(Area,2,sd)/sqrt(10)
```

A.8.4 Section 6.5

The code for producing the results with $p = 100$ in Table 6.1:

```
library(equSA)
library(PRROC)

dim<-100
area <-NULL
for(M in 1:10){

   #### p-learning #####
   x<-DAGsim(n=200, p=dim, type="AR(2)", p.binary=50)
   A<-x$moral.matrix

   edge1 <-NULL; edge0 <-NULL; m<-0
   for(i in 1:dim){
      for(j in 1:dim){
         if(i!=j){
            m<-m+1
            if(A[i,j]==1) edge1 <-rbind(edge1, c(i,j,m))
```

```
          else  edge0<-rbind(edge0,c(i,j,m))
      }
    }
  }
  g<-plearn.moral(x$data,alpha1=0.05,alpha2=0.05)

  s<-g$score
  s1<-s[edge1[,3],3]
  s0<-s[edge0[,3],3]
  roc<-roc.curve(s1,s0,curve=TRUE)
  area<-c(area,roc$auc)
}
```

A.8.5 Section 7.4.1

The code for producing the results of the structure "cluster" in Table 7.1:

```
library(equSA)
library(JGL)
library(PRROC)
library(flux)
library(MASS)


epsilon <-0.1; dim<-200; samsize<-100
double_random<-function(m,smin,smax){
    a<-sample(c(-1,1),1);
    b<-runif(m,smin,smax);
    a*b;
}


AREA<-NULL
for(M in 1:10){

    ##### Data generation: condition 1 #########
    x1<-GauSim(n=samsize, p=dim, graph="cluster")
    z1<-x1$data;   A<-x1$theta

    edge11<-NULL; edge10<-NULL; m<-0
    for(i in 1:(dim-1))
      for(j in (i+1):dim){
        m<-m+1
        if(A[i,j]==1) edge11<-rbind(edge11,c(m,i,j))
            else  edge10<-rbind(edge10,c(m,i,j))
      }
    theta1 <-solve(x1$sigma)
```

```
######### condition 2 ########
ml<-floor(nrow(edge11)*0.05)
s1<-sample(1:nrow(edge11),ml)
s2<-sample(1:nrow(edge10),ml)
edge21<-rbind(edge11[-s1,],edge10[s2,])
edge20<-rbind(edge11[s1,],edge10[-s2,])

theta2<-theta1;
  for(i in 1:ml){
     theta2[edge11[s1[i],2],edge11[s1[i],3]]<-0;
     theta2[edge11[s1[i],3],edge11[s1[i],2]]<-0;
     r<-double_random(1,0.2,0.4);
     theta2[edge10[s2[i],2],edge10[s2[i],3]]<-r;
     theta2[edge10[s2[i],3],edge10[s2[i],2]]<-r;
     }
f<-min(eigen(theta2)$values)
if(f<=epsilon) theta2<-theta2+diag(-f+epsilon,dim)
z2<-mvrnorm(n=samsize, rep(0,dim), solve(theta2))

############ condition 3 ##########
ml<-floor(nrow(edge21)*0.05)
s1<-sample(1:nrow(edge21),ml)
s2<-sample(1:nrow(edge20),ml)
edge31<-rbind(edge21[-s1,],edge20[s2,])
edge30<-rbind(edge21[s1,],edge20[-s2,])

theta3<-theta2;
for(i in 1:ml){
     theta3[edge21[s1[i],2],edge21[s1[i],3]]<-0;
     theta3[edge21[s1[i],3],edge21[s1[i],2]]<-0;
     r<-double_random(1,0.2,0.4);
     theta3[edge20[s2[i],2],edge20[s2[i],3]]<-r;
     theta3[edge20[s2[i],3],edge20[s2[i],2]]<-r;
     }
f<-min(eigen(theta3)$values)
if(f<=epsilon) theta3<-theta3+diag(-f+epsilon,dim)
z3<-mvrnorm(n=samsize, rep(0,dim), solve(theta3))

############ condition 4 ###########
ml<-floor(nrow(edge31)*0.05)
s1<-sample(1:nrow(edge31),ml)
s2<-sample(1:nrow(edge30),ml)
edge41<-rbind(edge31[-s1,],edge30[s2,])
edge40<-rbind(edge31[s1,],edge30[-s2,])
```

```
theta4 <-theta3 ;
for ( i  in  1:ml){
    theta4 [ edge31 [ s1 [ i ] ,2 ] , edge31 [ s1 [ i ],3]] < -0;
    theta4 [ edge31 [ s1 [ i ] ,3 ] , edge31 [ s1 [ i ],2]] < -0;
    r <-double_random ( 1 ,0.2 ,0.4);
    theta4 [ edge30 [ s2 [ i ] ,2 ] , edge30 [ s2 [ i ],3]] < -r ;
    theta4 [ edge30 [ s2 [ i ] ,3 ] , edge30 [ s2 [ i ],2]] < -r ;
    }
f <-min ( eigen ( theta4 ) $values )
if ( f <=epsilon )  theta4 <-theta4+diag (-f+epsilon ,dim )
z4 <-mvrnorm ( n=samsize ,  rep ( 0 ,dim ),  solve ( theta4 ))

A1 <-diag ( 0 ,dim );  A1 [ edge11 [ ,2:3]] < -1;   A1 <-A1+t ( A1 );
A2 <-diag ( 0 ,dim );  A2 [ edge21 [ ,2:3]] < -1;   A2 <-A2+t ( A2 );
A3 <-diag ( 0 ,dim );  A3 [ edge31 [ ,2:3]] < -1;   A3 <-A3+t ( A3 );
A4 <-diag ( 0 ,dim );  A4 [ edge41 [ ,2:3]] < -1;   A4 <-A4+t ( A4 );

data_all <-vector ( " list " ,4 )
data_all [[1]] < -z1 ;  data_all [[2]] < -z2 ;
data_all [[3]] < -z3 ;  data_all [[4]] < -z4 ;

area <-NULL
#################### FHBIA  ###################
g <-JGGM( data_all ,ALPHA1=0.05 ,ALPHA2=0.05 ,
        parallel=TRUE, nCPUs=4 )

### joint analysis ###
SJ <-g$score.joint ;
s1 <-SJ[ edge11 [ ,1 ] ,3]; e1 <-SJ[ edge10 [ ,1 ],3];
s2 <-SJ[ edge21 [ ,1 ] ,4]; e2 <-SJ[ edge20 [ ,1 ],4];
s3 <-SJ[ edge31 [ ,1 ] ,5]; e3 <-SJ[ edge30 [ ,1 ],5];
s4 <-SJ[ edge41 [ ,1 ] ,6]; e4 <-SJ[ edge40 [ ,1 ],6];
sall <-(-1)*c ( s1 ,s2 ,s3 ,s4 );  eall <-(-1)*c ( e1 ,e2 ,e3 ,e4 )
prj <-pr.curve ( sall ,eall ,curve=TRUE)
area <-c ( area ,prj$auc.integral )

##### separate analysis #####
SP <-g$score.sep ;
s1 <-SP[ edge11 [ ,1 ] ,3]; e1 <-SP[ edge10 [ ,1 ],3];
s2 <-SP[ edge21 [ ,1 ] ,4]; e2 <-SP[ edge20 [ ,1 ],4];
s3 <-SP[ edge31 [ ,1 ] ,5]; e3 <-SP[ edge30 [ ,1 ],5];
s4 <-SP[ edge41 [ ,1 ] ,6]; e4 <-SP[ edge40 [ ,1 ],6];
sall <-(-1)*c ( s1 ,s2 ,s3 ,s4 );  eall <-(-1)*c ( e1 ,e2 ,e3 ,e4 )
prs <-pr.curve ( sall ,eall ,curve=TRUE)
area <-c ( area ,  prs$auc.integral )
```

```
################ JGL    #############################
sig2 <-c(0.05,0.1,0.2)
sig1 <-seq(0.025,0.725,len=20)

##### fused lasso ######
for(lam2 in sig2){
  R1<-NULL
  for(lam1 in sig1){
    g2<-JGL(data_all,penalty="fused",lambda1=lam1,
            lambda2=lam2, return.whole.theta=TRUE)

    B1<-g2$theta[[1]];  B2<-g2$theta[[2]];
    B3<-g2$theta[[3]];  B4<-g2$theta[[4]];
    B1[which(B1!=0)]<-1;  B2[which(B2!=0)]<-1;
    B3[which(B3!=0)]<-1;  B4[which(B4!=0)]<-1
    B1<-B1-diag(1,dim);  B2<-B2-diag(1,dim);
    B3<-B3-diag(1,dim);  B4<-B4-diag(1,dim);
    l1<-length(which(A1+B1==2));  l2<-length
        (which(A2+B2==2));
    l3<-length(which(A3+B3==2));  l4<-length
        (which(A4+B4==2));
    precision <-1.0*(l1+l2+l3+l4)/sum(B1+B2+B3+B4)
    recall <-1.0*(l1+l2+l3+l4)/sum(A1+A2+A3+A4)
    R1<-rbind(R1,c(recall,precision))
  }
R1<-rbind(R1,c(0,1));  R1<-rbind(R1,c(1,0))
auc.fused<-auc(R1[,1],R1[,2])
area<-c(area,auc.fused)
}

### group Lasso #####
for(lam2 in sig2){
  R2<-NULL
  for(lam1 in sig1){
    g3<-JGL(data_all, penalty="group", lambda1=lam1,
            lambda2=0.05, return.whole.theta=TRUE)

    B1<-g3$theta[[1]];  B2<-g3$theta[[2]];
    B3<-g3$theta[[3]];  B4<-g3$theta[[4]];
    B1[which(B1!=0)]<-1;  B2[which(B2!=0)]<-1;
    B3[which(B3!=0)]<-1;  B4[which(B4!=0)]<-1
    B1<-B1-diag(1,dim);  B2<-B2-diag(1,dim);
    B3<-B3-diag(1,dim);  B4<-B4-diag(1,dim);
    l1<-length(which(A1+B1==2));  l2<-length
        (which(A2+B2==2));
```

```
    l3 <-length ( which (A3+B3==2));   l4 <-length
       ( which (A4+B4==2));
    precision <-1.0*(l1+l2+l3+l4 )/ sum (B1+B2+B3+B4)
    recall <-1.0*(l1+l2+l3+l4 )/ sum (A1+A2+A3+A4)
    R2<-rbind (R2,c (recall , precision ))
  }
  R2<-rbind (R2,c (0 ,1));   R2<-rbind (R2,c (1 ,0))
  auc . group <-auc (R2[ ,1] ,R2[ ,2])
  area <-c ( area , auc . group )
 }
 AREA<-rbind (AREA,  area )
}
apply (AREA,2 ,mean )
apply (AREA,2 ,sd )/ sqrt (10)
```

A.8.6   Section 8.4.1

The code for producing the results of double regression in Table 8.1:

```
library ( parallel )
library ( huge )
library ( GeneralisedCovarianceMeasure )
library (PRROC)

area <-NULL
for ( i  in  1:50){
   ####### Data generation #######################
   num< -200;  dim< -30
   x1<-runif (num, -1 ,1);   x2<-rnorm (num, -1 ,1);
   x3<-rnorm (num,0 ,1);   x4<-rnorm (num,0 ,1);
   x5<-rnorm (num,0 ,1);    x6<-runif (num, -1 ,1);
   x7<-runif (num, -1 ,1);

   x2<-6*cos (x1)+x2 ;   x3<-5*sin (x1)+x2+x3 ;
   x4<-5*cos (x3*x6)+3*x3+3*x6+x4;  x5 <-0.05*(x2+x6)^3+x5 ;
   x7<-6*cos (0.2*( x3+log ( abs (5*x5 )+1)))+x7 ;

   X<-cbind (x1 ,x2 ,x3 ,x4 ,x5 ,x6 ,x7 )
   for ( i  in  8:dim)  X<-cbind (X, rnorm (num,0 ,1))
   write (round ( t(X) ,5),  ncol=ncol(X),  file ="sim . data ")

   ####### Variable sure  screening #############
   HZSIS<-function (X){
      n<-nrow (X)
      p<-ncol (X)
```

```
    beta<-exp(1.0/(p+4)*log(n*(2*p+1)/4.0))/sqrt(2)
    m<-apply(X*X,1,sum)
    s2<-sum(exp(-0.5*beta*beta/(1+beta*beta)*m))
    s1<-0
    for(i in 1:n){
      A<-t(matrix(rep(X[i,],n), ncol=n))
      D<-A-X
      d<-apply(D*D,1,sum)
      s1<-s1+sum(exp(-0.5*beta*beta*d))
      }
    hz<-s1/n-2*exp(-0.5*p*log(1+beta*beta))*s2
        +n*exp(-0.5*p*log(1+2*beta*beta))
    a<-1+2*beta*beta
    w<-(1+beta^2)*(1+3*beta^2)
    ave<-1-a^(-p/2)*(1+p*beta^2/a+p*(p+2)*beta^4/(2*a^2))
    SS<-2*(1+4*beta*beta)^(-p/2)+2*a^(-p)*(1+2*p*beta^4/
        (a^2)+0.75*p*(p+2)*beta^8/(a^4))
    SS<-SS-4*w^(-p/2)*(1+1.5*p*beta^4/w
        +0.5*p*(p+2)*beta^8/(w^2))
    sigmasq<-log(1+SS/ave^2)
    mu<-log(ave)-0.5*sigmasq
    z<-(log(hz)-mu)/sqrt(sigmasq)
    return(z)
  }

neighborsize<-26
#X<-as.matrix(read.table("sim.data"))
Z<-huge.npn(X)
P<-ncol(Z)

myneighbor<-function(w){
    score<-1:P
    for(k in 1:P){
        d<-cbind(w,Z[,k])
        score[k]<-HZSIS(d)
        }
    U<-cbind(1:P, score)
    U<-U[order(-U[,2]),1:2]
    sel<-U[1:neighborsize,1]
    return(sel)
  }

D<-list()
for(i in 1:P){
  D[[i]]<-Z[,i]
```

```
      }

   system.time({d<-mclapply(D, myneighbor, mc.cores=10)
              S2<-do.call("rbind",d)
              })
   write(t(S2), ncol=ncol(S2), file="neighbor.txt")

   ######## Multiple hypothesis tests #############
   #X<-as.matrix(read.table("sim.data"))
   #S2<-as.matrix(read.table("neighbor.txt"))
   neisze<-6

   mytest<-function(a){
        i<-a[1]; j<-a[2];
        n1<-S2[i,2:neisze]; n2<-S2[j,2:neisze];
        nei<-setdiff(unique(c(n1,n2)),c(i,j));
        q<-gcm.test(X[,i], X[,j], X[,nei]);
        pv<-c(i,j,q$test.statistic, q$p.value);
        return(pv)
     }

   D<-list()
   k<-0
   for(i in 2:P)
     for(j in 1:(i-1)){
          k<-k+1;
          D[[k]]<-c(i,j);
     }
   system.time({d<-mclapply(D, mytest, mc.cores=10)
                pvalue<-do.call("rbind",d)
                })

   fdr<-p.adjust(pvalue[,4], method="fdr")
   pvalue<-cbind(pvalue,fdr)
   sel<-pvalue[pvalue[,5]<0.05,]

   ### define a set w for the true edges of the network
   w<-c(1,2,3,6,8,9,12,13,14,15,18,20)
   s1<-1-pvalue[w,5]; s0<-1-pvalue[-w,5]
   pr<-pr.curve(s1,s0)
   area<-c(area,pr$auc.integral)
}
```

A.8.7 Section 9.5

The code for producing the results of linear regression in Table 9.1:

```
library (mvtnorm)
library (SIS)
library (equSA)

n<-200; p<-500; sigma_eps_true <- 1
CI_hat <-NULL; beta_hat <-NULL; pvalue_hat <-NULL
for(iter in 1:100){

  mu_x_true <- rep(0,p)
  beta_0 <-rep(1,n)
  beta_true <- rep(0,p)
  beta_true[1:5] <- c(2,2.25,3,3.25,3.75)

  C <- diag(1,p)
  for(i in 1:(p-1)) C[i,i+1]<-C[i+1,i]<-0.5
  for(i in 1:(p-2)) C[i,i+2]<-C[i+2,i]<-0.25
  sigma<-solve(C)

  sigma2<-sigma;
  for(i in 1:p)
  {
    for(j in 1:p)
    {
      if(sigma[i,j]<=sigma[j,i]) {sigma2[i,j]=sigma[j,i]}
      else {sigma2[i,j]=sigma[i,j]};
    }
  }
  x <- rmvnorm(n,mu_x_true,sigma2)
  eps <- rnorm(n,0,sigma_eps_true)
  y <- beta_0 +x%*%beta_true+eps

  #### SIS-MCP for variable selection ###############
  fit_SIS <- SIS(x,y,family = 'gaussian',penalty = 'MCP',
                  tune='bic', seed=2*iter*sqrt(100))
  penal_sel <- fit_SIS$ix

  #### MNR method for high-dimensional inference #####
  Res <- equSAR(x,ALPHA1=0.1,ALPHA2=0.05)
  A1<-Res$Adj

  CI <- NULL; beta <- NULL; pvalue <- NULL;
  for(k in 1:p){
```

```
neibor <- which(A1[k,]==1)
x_sel_name <- unique(c(neibor,penal_sel,k))
x_sel <- x[,match(x_sel_name,1:p)]
colnames(x_sel) <- x_sel_name
#dim(x_sel)
fit_lm2 <- lm(y~x_sel)
CI<-rbind(CI,confint(fit_lm2)[which(x_sel_name==k)
    +1,])
beta<-c(beta,fit_lm2$coefficients
    [which(x_sel_name==k)+1])
pvalue<-c(pvalue,summary(fit_lm2)$coefficients
            [which(x_sel_name ==k)+1,4])
}
CI_hat <- cbind(CI_hat,CI)
beta_hat <- cbind(beta_hat,beta)
pvalue_hat <- cbind(pvalue_hat,pvalue)
}
```

Bibliography

[1] M. Akkiprik, Y. Feng, H. Wang, K. Chen, L. Hu, A. Sahin, S. Krishnamurthy, A. Özer, X. Hao, and W. Zhang. Multifunctional roles of insulin-like growth factor binding protein 5 in breast cancer. *Breast Cancer Research : BCR*, 10:212–212, 2008.

[2] C.F. Aliferis, A. Statnikov, I. Tsamardinos, S. Mani, and X. D. Koutsoukos. Local causal and Markov blanket induction for causal discovery and feature selection for classification, part I: Algorithms and empirical evaluation. *Journal of Machine Learning Research*, 11:171–234, 2010.

[3] G. Allen and Z. Liu. A local poisson graphical model for inferring networks from sequencing data. *IEEE Transactions on NanoBioscience*, 12(3):189–198, 2013.

[4] S. Anders and W. Huber. Differential expression analysis for sequence count data. *Genome Biology*, 11:R106–R106, 2010.

[5] F. Bachoc, D. Preinerstorfer, and L. Steinberger. Uniformly valid confidence intervals post-model-selection. *The Annals of Statistics*, 48(1):440–463, 2020.

[6] O. Banerjee, L. El Ghaoui, and A. d'Aspremont. Model selection through sparse maximum likelihood estimation for multivariate Gaussian or binary data. *Journal of Machine Learning Research*, 9(15):485–516, 2008.

[7] A.L. Barabási and R. Albert. Emergence of scaling in random networks. *Science*, 286(5439):509–512, 1999.

[8] J. Barretina, G. Caponigro, N. Stransky, K. Venkatesan, A.A. Margolin, S. Kim, C.J. Wilson, J. Lehár, G.V. Kryukov, D. Sonkin, A. Reddy, M. Liu, L. Murray, M.F. Berger, J.E. Monahan, P. Morais, J. Meltzer, A. Korejwa, J. Jané-Valbuena, F.A. Mapa, J. Thibault, E. Bric-Furlong, P. Raman, A. Shipwayn, I.H. Engels, J. Cheng, G.K. Yu, J. Yu, P. Aspesi, M. de Silva, K. Jagtap, M.D. Jones, L. Wang, C. Hatton, E. Palescandolo, S. Gupta, S. Mahan, C. Sougnez, R.C. Onofrio, T. Liefeld, L. MacConaill, W. Winckler, M. Reich, N. Li, J.P. Mesirov, S.B. Gabriel, G. Getz, K. Ardlie, V. Chan, V.E. Myer, B.L. Weber, J. Porter, M. Warmuth, P. Finan, J.L. Harris, M. Meyerson, T.R. Golub, M.P. Morrissey, W.R. Sellers, R. Schlegel, and L.A. Garraway. The cancer cell line encyclopedia enables predictive modeling of anticancer drug sensitivity. *Nature*, 483:603–607, 2012.

[9] A. Bellot and M. van der Schaar. Conditional independence testing using generative adversarial networks. In: H.M. Wallach, H. Larochelle, A. Beygelzimer, F. d'Alché-Buc, E.B. Fox, and R. Garnett, editors, *Advances in Neural Information Processing Systems 32: Annual Conference on Neural Information Processing Systems 2019, NeurIPS 2019*, Vancouver, BC, Canada, December 8–14, 2019, pp. 2199–2208, 2019.

[10] Y. Benjamini, A.M. Krieger, and D. Yekutieli. Adaptive linear step-up procedures that control the false discovery rate. *Biometrika*, 93:491–507, 2006.

[11] R.A. Berk, L.D. Brown, A. Buja, K. Zhang, and L.H. Zhao. Valid post-selection inference. *Annals of Statistics*, 41:802–837, 2013.

[12] T. Berrett, Y. Wang, R. Barber, and R. Samworth. The conditional permutation test for independence while controlling for confounders. *Journal of the Royal Statistical Society: Series B (Statistical Methodology)*, 82(1):175–197, 2020.

[13] P. Bertheau, J. Lehmann-Che, M. Varna, A. Dumay, B. Poirot, R. Porcher, E. Turpin, L.F. Plassa, A. de Roquancourt, E. Bourstyn, P. de Cremoux, A. Janin, S. Giacchetti, M. Espie, and H. de The. p53 in breast cancer subtypes and new insights into response to chemotherapy. *Breast*, Suppl 2:S27–S29, 2013.

[14] J. Besag. Spatial interaction and the statistical analysis of lattice systems. *Journal of the Royal Statistical Society, Series B*, 36(2):192–236, 1974.

[15] T. Hellem Bø, B. Dysvik, and I. Jonassen. Lsimpute: Accurate estimation of missing values in microarray data with least squares methods. *Nucleic Acids Research*, 32(3):e34, 2004.

[16] M.B. Brown. 400: A method for combining non-independent, one-sided tests of significance. *Biometrics*, 31:987, 1975.

[17] P. Bühlmann. Statistical significance in high-dimensional linear models. *Bernoulli*, 19:1212–1242, 2013.

[18] P. Bühlmann, M. Kalisch, and M.H. Maathuis. Variable selection in high-dimensional linear models: Patially faithful distributions and the pc-simple algorithm. *Biometrika*, 97(2):261–278, 2010.

[19] T. Cai, H. Li, W. Liu, and J. Xie. Covariate-adjusted precision matrix estimation with an application in genetical genomics. *Biometrika*, 100(1):139–156, 2013.

[20] G. Celeux, D. Chauveau, and J. Diebolt. Stochastic versions of the em algorithm: An experimental study in the mixture case. *Journal of Statistical Computation and Simulation*, 55:287–314, 1996.

[21] A. Chatterjee and S.N. Lahiri. Rates of convergence of the adaptive Lasso estimators to the oracle distribution and higher order refinements by the bootstrap. *Annals of Statistics*, 41:1232–1259, 2013.

[22] S. Chen, D.M. Witten, and A. Shojaie. Selection and estimation for mixed graphical models. *Biometrika*, 102(1):47–64, 2015.

[23] J. Cheng, T. Li, E. Levina, and J. Zhu. High-dimensional mixed graphical models. *Journal of Computational and Graphical Statistics*, 26:367–378, 2017.

[24] X. Cheng, H. Wang, L. Zhu, W. Zhong, and H. Zhou. Moder-free sure independent screening procedures. *R Package*, https://rdrr.io/cran/MFSIS/src/R/MFSIS.R, 2022.

[25] D.M. Chickering. Learnin Bayesian networks is np-complete. In: D. Fisher and H.-J. Lenz, editors, *Learning from Data: Artificial Intelligence and Statistics V*, pp. 121–130. Springer-Verlag, New York, 1996.

[26] D.M. Chickering. Optimal structure identification with greedy search. *Journal of Machine Learning Research*, 3:507–554, 2002.

[27] J. Choi, R.S. Chapkin, and Y. Ni. Bayesian causal structural learning with zero-inflated poisson Bayesian networks. In: H. Larochelle, M. Ranzato, R. Hadsell, M.-F. Balcan, and H.-T. Lin, editors, *Advances in Neural Information Processing Systems 33: Annual Conference on Neural Information Processing Systems 2020, NeurIPS 2020*, December 6–12, 2020, virtual, 2020.

[28] H. Chun, X. Zhang, and H. Zhao. Gene regulation network inference with joint sparse Gaussian graphical models. *Journal of Computational and Graphical Statistics*, 24:954–974, 2015.

[29] D. Colombo and M.H. Maathuis. Order-independent constraint-based causal structure learning. *Journal of Machine Learning Research*, 15:3741–3782, 2014.

[30] G.F. Cooper and E. Herskovits. A Bayesian method for the induction of probabilistic networks from data. *Machine Learning*, 9:309–347, 1992.

[31] G.F. Cooper and C. Yoo. Causal discovery from a mixture of experimental and observational data. In: K.B. Laskey and H. Prade, editors, *UAI '99: Proceedings of the Fifteenth Conference on Uncertainty in Artificial Intelligence*, Stockholm, Sweden, July 30 to August 1, 1999, pp. 116–125. Morgan Kaufmann, 1999.

[32] H. Cui, R. Li, and W. Zhong. Model-free feature screening for ultrahigh dimensional discriminant analysis. *Journal of the American Statistical Association*, 110(510):630–641, 2015.

[33] P. Danaher. Performs the joint graphical Lasso for sparse inverse covariance estimation on multiple classes, R package version 2.3.1, 2018.

[34] P.J. Danaher, P. Wang, and D. Witten. The joint graphical Lasso for inverse covariance estimation across multiple classes. *Journal of the Royal Statistical Society: Series B (Statistical Methodology)*, 76(2):373–397, 2014.

[35] A.P. Dempster, N.M. Laird, and D.B. Rubin. Maximum likelihood estimation from incomplete data via the em algorithm (with discussion). *Journal of the Royal Statistical Society, Series B*, 39:1–38, 1977.

[36] B. Deverett and Ch. Kemp. Learning deterministic causal networks from observational data. In: N. Miyake, D. Peebles, and R.P. Cooper, editors, *Proceedings of the 34th Annual Meeting of the Cognitive Science Society, CogSci 2012*, Sapporo, Japan, August 1-4, 2012, pp. 288–293, cognitivesciencesociety.org, 2012.

[37] R. Dezeure, P. Bühlmann, L. Meier, and N. Meinshausen. High-dimensional inference: Confidence intervals, p-values and r-software HDI. *Statistical Science*, 30(4):533–558, 2015.

[38] A. Dobra and A. Lenkoski. Copula Gaussian graphical models and their application to modeling functional disability data. *The Annals of Applied Statistics*, 5:969–993, 2011.

[39] G. Doran, K. Muandet, K. Zhang, and B. Schölkopf. A permutation-based kernel conditional independence test. In: N.L. Zhang and J. Tian, editors, *Proceedings of the Thirtieth Conference on Uncertainty in Artificial Intelligence, UAI 2014*, Quebec City, Quebec, Canada, July 23–27, 2014, pp. 132–141. AUAI Press, 2014.

[40] S. Dorman, K. Baranova, J. Knoll, B. Urquhart, G. Mariani, M. Carcangiu, and P. Rogan. Genomic signatures for paclitaxel and gemcitabine resistance in breast cancer drived by machine learning. *Molecular Oncology*, 10:85–100, 2016.

[41] B. Efron. Large-scale simultaneous hypothesis testing: The choice of a null hypothesis. *Journal of the American Statistical Association*, 99:96–104, 2004.

[42] B. Ellis and W.H. Wong. Learning causal Bayesian network structures from experimental data. *Journal of the American Statistical Association*, 103:778–789, 2008.

[43] J. Fan and R. Li. Variable selection via nonconcave penalized likelihood and its oracle properties. *Journal of the American Statistical Association*, 96:1348–1360, 2001.

[44] J. Fan, H. Liu, Y. Ning, and H. Zou. High dimensional semiparametric latent graphical model for mixed data. *Journal of the Royal Statistical Society, Series B*, 79(2):405–421, 2017.

[45] J. Fan and J. Lv. Sure independence screening for ultrahigh dimensional feature space. *Journal of the Royal Statistical Society: Series B (Statistical Methodology)*, 70(5):849–911, 2008.

[46] J. Fan and R. Song. Sure independence screening in generalized linear models with np-dimensionality. *The Annals of Statistics*, 38(6):3567–3604, 2010.

[47] B. Fellinghauer, P. Bühlmann, M. Ryffel, M. von Rhein, and J.D. Reinhardt. Stable graphical model estimation with random forests for discrete, continuous, and mixed variables. *Computational Statistics & Data Analysis*, 64:132–152, 2013.

[48] W. Fithian, D.L. Sun, and J.E. Taylor. Optimal inference after model selection. arXiv: Statistics Theory, 2014.

[49] J. Friedman, T. Hastie, and R. Tibshirani. Sparse inverse covariance estimation with the graphical Lasso. *Biostatistics*, 9(3):432–441, 2008.

[50] J.H. Friedman, T.J. Hastie, H. Hofling, and R. Tibshirani. Pathwise coordinate optimization. *The Annals of Applied Statistics*, 1:302–332, 2007.

[51] N. Friedman and D. Koller. Being Bayesian about network structure: A Bayesian approach to structure sidcovery in Bayesian networks. *Machine Learning*, 50:95–125, 2003.

[52] N. Friedman, I. Nachman, and D. Pe'er. Learning Bayesian network structure from massive datasets: The "sparse candidate" algorithm. In: K.B. Laskey and H. Prade, editors, *UAI '99: Proceedings of the Fifteenth Conference on Uncertainty in Artificial Intelligence*, Stockholm, Sweden, July 30 to August 1, 1999, pp. 206–215. Morgan Kaufmann, 1999.

[53] M. Gallopin, A. Rau, and F. Jaffrézic. A hierarchical poisson log-normal model for network inference from RNA sequencing data. *PLoS One*, 8(10):e77503, 2013.

[54] L. Gan, N.N. Narisetty, and F. Liang. Bayesian regularization for graphical models with unequal shrinkage. *Journal of the American Statistical Association*, 114:1218–1231, 2019.

[55] A.P. Gasch, P.T. Spellman, C.M. Kao, O. Carmel-Harel, M.B. Eisen, G. Storz, D. Botstein, and P.O. Brown. Genomic expression programs in the response of yeast cells to environmental changes. *Molecular Biology of the Cell*, 11(12):4241–4257, 2000.

[56] S. Geman and D. Geman. Stochastic relaxation, Gibbs distributions, and the Bayesian restoration of images. *IEEE Transactions on Pattern Analysis and Machine Intelligence*, PAMI-6:721–741, 1984.

[57] J. Guo, E. Levina, G. Michailidis, and J. Zhu. Joint estimation of multiple graphical models. *Biometrika*, 98(1):1–15, 2011.

[58] K.E. Hadley and D.T. Hendricks. Use of nqo1 status as a selective biomarker for oesophageal squamous cell carcinomas with greater sensitivity to 17-aag. *BMC Cancer*, 14:1–8, 2014.

[59] R. Haque, S.A. Ahmed, G. Inzhakova, J.M. Shi, C.C. Avila, J.A. Polikoff, L. Bernstein, S.M. Enger, and M.F Press. Impact of breast cancer subtypes and treatment on survival: An analysis spanning two decades. *Cancer Epidemiology, Biomarkers & Prevention*, 21:1848–1855, 2012.

[60] T.J. Hastie, R. Tibshirani, and J.H. Friedman. *The Elements of Statistical Learning*, (2nd edition). Springer, New York, 2009.

[61] T.J. Hastie, R. Tibshirani, and M.J. Wainwright. *Statistical Learning with Sparsity: The Lasso and Generalizations*. CRC Press, Boca Raton, FL, 2015.

[62] W.K. Hastings. Monte carlo sampling methods using Markov chain and their applications. *Biometrika*, 57:97–109, 1970.

[63] S. He. spaceExt: Undirected graph inference with missing data. R package version 1.0, 2011.

[64] Y. He, T. Jiang, J. Wen, and G. Xu. Likelihood ratio test in multivariate linear regression: From low to high dimension. *Statistica Sinica*, 31:1215–1238, 2021.

[65] P.D. Hoff. Extending the rank likelihood for semiparametric copula estimation. *The Annals of Applied Statistics*, 1:265–283, 2006.

[66] S. Holm. A simple sequentially rejective multiple test procedure. *Scandinavian Journal of Statistics*, 6:65–70, 1979.

[67] H. Holzmann, A. Munk, and T. Gneiting. Identifiability of Finite Mixtures of Elliptical Distributions. *Scandinavian Journal of Statistics*, 33:753–763, 2006.

[68] P.O. Hoyer, A. Hyvärinen, R. Scheines, P. Spirtes, J.D. Ramsey, G. Lacerda, and S. Shimizu. Causal discovery of linear acyclic models with arbitrary distributions. In: D.A. McAllester and P. Myllymäki, editors, *UAI 2008, Proceedings of the 24th Conference in Uncertainty in Artificial Intelligence*, Helsinki, Finland, July 9–12, 2008, pp. 282–289. AUAI Press, 2008.

[69] M. Humbert, V. Halter, D. Shan, J. Laedrach, E.O. Leibundgut, G. Baerlocher, A. Tobler, M. Fey, and M. Tschan. Deregulated expression of kruppel-like factors in acute myeloid leukemia. *Leukemia Research*, 35(7):909–913, 2011.

[70] D.I. Inouye, P. Ravikumar, and I.S. Dhillon. Square root graphical models: Multivariate generalizations of univariate exponential families that permit positive dependencies. In: M.-F. Balcan and K.Q. Weinberger, editors, *Proceedings of the 33nd International Conference on Machine Learning, ICML 2016*, New York City, June 19–24, 2016, volume 48 of *JMLR Workshop and Conference Proceedings*, pp. 2445–2453. JMLR.org, 2016.

[71] A. Javanmard and A. Montanari. Confidence intervals and hypothesis testing for high-dimensional regression. *Journal of Machine Learning Research*, 15:2869–2909, 2014.

[72] F.V. Jensen and T.D. Nielsen. *Bayesian Networks and Decision Graphs*. Springer, New York, 2007.

[73] B. Jia and F. Liang. Learning gene regulatory networks with high-dimensional heterogeneous data. In: Y. Zhao and D.-G. Chen, editors, *Frontiers of Biostatistics and Bioinformatics*, pp. 305–327. Springer, New York, 2018.

[74] B. Jia, S. Xu, G. Xiao, V. Lamba, and F. Liang. Learning gene regulatory networks from next generation sequencing data. *Biometrics*, 73(4):1221–1230, 2017.

[75] B. Jia and F. Liang. Joint estimation of multiple mixed graphical models for pan-cancer network analysis. *Stat*, 9:e271, 2020.

[76] B. Jia, F. Liang, R. Shi, and S. Xu. Equsa: Learning high-dimensional graphical models. *R package*, 2019.

[77] B. Jia, F. Liang, and TEDDY Study Group. Fast hybrid Bayesian integrative learning of multiple gene regulatory networks for type 1 diabetes. *Biostatistics*, 22(2):233–249, 2021.

[78] M. Kalisch and P. Bühlmann. Estimating high-dimensional directed acyclic graphs with the pc-algorithm. *Journal of Machine Learning Research*, 8:613–636, 2007.

[79] K. Kaneko, S. Ishigami, Y. Kijima, Y. Funasako, M. Hirata, H. Okumura, H. Shinchi, C. Koriyama, S. Ueno, H. Yoshinaka, and S. Natsugoe. Clinical implication of HLA class I expression in breast cancer. *BMC Cancer*, 11:454–454, 2011.

[80] U.B. Kjaerulff and A.L. Madsen. *Bayesian Networks and Influence Diagrams*. Springer, New York, 2010.

[81] E. Kolaczyk. *Statistical Analysis of Network Data*. Springer, New York, 2009.

[82] J. Kost and M. McDermott. Combining dependent p-values. *Statistics & Probability Letters*, 60:183–190, 2002.

[83] A.K. Kuchibhotla, J.E. Kolassa, and T.A. Kuffner. Post-selection inference. *Annual Review of Statistics and Its Application*, 9:505–527, 2022.

[84] W. Lam and F. Bacchus. Learning Bayesian belief networks: An approach based on the MDL principle. *Computational Intelligence*, 10:269–293, 1994.

[85] S. Lauritzen. *Graphical Models*. Oxford University Press, Oxford, 1996.

[86] H. Lee, M. Hanibuchi, S.-J. Lim, H. Yu, M. Kim, J. He, R. Langley, F. Lehembre, U. Regenass, and I. Fidler. Treatment of experimental human breast cancer and lung cancer brain metastases in mice by macitentan, a dual antagonist of endothelin receptors, combined with paclitaxel. *NeuroOncology*, 18:486–496, 2016.

[87] J. Lee, D.L. Sun, Y. Sun, and J.E. Taylor. Exact post-selection inference, with application to the Lasso. *Annals of Statistics*, 44:907–927, 2016.

[88] J.D Lee and T.J. Hastie. Learning the structure of mixed graphical models. *Journal of Computational and Graphical Statistics*, 24(1):230–253, 2015.

[89] S. Lee, F. Liang, L. Cai, and G. Xiao. A two-stage approach of gene network analysis for high-dimensional heterogeneous data. *Biostatistics*, 19(2):216–232, 2018.

[90] C. Li and X. Fan. On nonparametric conditional independence tests for continuous variables. *Wiley Interdisciplinary Reviews: Computational Statistics*, 12:e1489, 2019.

[91] K.-H. Li. Imputation using Markov chains. *Journal of Statistical Computation and Simulation*, 30:57–79, 1988.

[92] F. Liang, B. Jia, J. Xue, Q. Li, and Y. Luo. An imputation-consistency approach for high-dimensional missing data problems and beyond. *Journal of the Royal Statistical Society, Series B*, 80(5):899–926, 2018.

[93] F. Liang, Q. Song, and P. Qiu. An equivalent measure of partial correlation coefficients for high dimensional Gaussian graphical models. *Journal of the American Statistical Association*, 110:1248–1265, 2015.

[94] [Liang et al.] F. Liang, J. Xue, and B. Jia. Markov neighborhood regression for high-dimensional inference. *Journal of the American Statistical Association*, 117(539):1200–1214, 2022.

[95] F. Liang and J. Zhang. Estimating the false discovery rate using the stochastic approximation algorithm. *Biometrika*, 95:961–977, 2008.

[96] F. Liang, Q. Li, and L. Zhou. Bayesian neural networks for selection of drug sensitive genes. *Journal of the American Statistical Association*, 113:955–972, 2018.

[97] F. Liang, C. Liu, and R.J. Carroll. *Advanced Markov Chain Monte Carlo Methods: Learning from Past Samples*. Wiley, Hoboken, NJ, 2010.

[98] F. Liang, Q. Song, and K. Yu. Bayesian subset modeling for high-dimensional generalized linear models. *Journal of the American Statistical Association*, 108:589–606, 2013.

[99] F. Liang and W.H. Wong. Evolutionary monte carlo: Applications to c_p model sampling and change point problem. *Statistica Sinica*, 10:317–342, 2000.

[100] F. Liang and J. Zhang. Learning Bayesian networks for discrete data. *Computational Statistics & Data Analysis*, 53:865–876, 2009.

[101] S. Liang and F. Liang. A double regression method for graphical modeling of high-dimensional nonlinear and non-Gaussian data. *Statistics and Its Interface*, in press, 2022.

[102] Z. Lin, T. Wang, C. Yang, and H. Zhao. On joint estimation of Gaussian graphical models for spatial and temporal data. *Biometrics*, 73(3):769–779, 2017.

[103] H. Liu, J.D. Lafferty, and L.A. Wasserman. The nonparanormal: semiparametric estimation of high dimensional undirected graphs. *Journal of Machine Learning Research*, 10:2295–2328, 2009.

[104] Y. Liu and B. Yu. Asymptotic properties of Lasso+mLS and Lasso+Ridge in sparse high-dimensional linear regression. *Electronic Journal of Statistics*, 7:3124–3169, 2013.

[105] P.-L. Loh and M.J. Wainwright. Regularized m-estimators with nonconvexity: Statistical and algorithmic theory for local optima. *Journal of Machine Learning Research*, 16:559–616, 2015.

[106] P.-L. Loh and M.J. Wainwright. Support recovery without incoherence: A case for nonconvex regularization. *Annals of Statistics*, 45:2455–2482, 2017.

[107] J.M. Louderbough and J.A. Schroeder. Understanding the dual nature of CD44 in breast cancer progression. *Molecular Cancer Research : MCR*, 9:1573–1586, 2011.

[108] S. Luo, R. Song, and D.M. Witten. Sure screening for Gaussian graphical models. ArXiv,abs/1407.7819, 2014.

[109] Z.-Q.T. Luo and P. Tseng. On the convergence of the coordinate descent method for convex differentiable minimization. *Journal of Optimization Theory and Applications*, 72:7–35, 1992.

[110] D. Margaritis. Learning Bayesian network structure models from data. PhD thesis, Carnegie-Mellon University, Pittsburgh, PA, 2003.

[111] D. Margaritis and S. Thrun. Bayesian network induction via local neighborhoods. In: S.A. Solla, T.K. Leen, and K.-R. Müller, editors, *Advances in Neural Information Processing Systems 12, [NIPS Conference*, Denver, CO, November 29 to December 4, 1999, pp. 505–511. The MIT Press, 1999.

[112] R. Mazumder and T. Hastie. The graphical Lasso: New insights and alternatives. *Electronic Journal of Statistics*, 6:2125–2149, 2012.

[113] C. Meek. Strong completeness and faithfulness in Bayesian networks. In: P. Besnard and S. Hanks, editors, *UAI '95: Proceedings of the Eleventh Annual Conference on Uncertainty in Artificial Intelligence*, Montreal, Quebec, Canada, August 18–20, 1995, pp. 411–418. Morgan Kaufmann, 1995.

[114] L. Meier, R. Dezeure, N. Meinshausen, M. Maechler, and P. Buehlmann. HDI: High-dimensional inference. *R Package*, https://CRAN.R–project.org/package=hdi, 2016.

[115] N. Meinshausen. Relaxed Lasso. *Computational Statistics & Data Analysis*, 52(1):374–393, 2007.

[116] N. Meinshausen, L. Meier, and P. Bühlmann. p-values for high-dimensional regression. *Journal of the American Statistical Association*, 104:1671–1681, 2009.

[117] N. Meinshausen. Group bound: Confidence intervals for groups of variables in sparse high dimensional regression without assumptions on the design. *Journal of the Royal Statistical Society: Series B (Statistical Methodology)*, 77:923–945, 2015.

[118] N. Meinshausen and P. Bühlmann. High-dimensional graphs and variable selection with the Lasso. *The Annals of Statistics*, 34(3):1436–1462, 2006.

[119] N. Metropolis, A.W. Rosenbluth, M.N. Rosenbluth, A.H. Teller, and E. Teller. Equation of state calculations by fast computing machines. *Journal of Chemical Physics*, 21:1087–1091, 1953.

[120] H. Mizuno, K. Kitada, K. Nakai, and A. Sarai. Prognoscan: A new database for meta-analysis of the prognostic value of genes. *BMC Medical Genomics*, 2:18–18, 2008.

[121] P. Müller. Alternatives to the Gibbs sampling scheme. Technical Report, Institute of Statistics and Decision Sciences, Duke University, 1992.

[122] P. Nandy, A. Hauser, and M.H. Maathuis. High-dimensional consistency in score-based and hybrid structure learning. arXiv:1507.02608v3, 2016.

[123] S. Nielsen. The stochastic em algorithm: Estimation and asymptotic results. *Bernoulli*, 6:457–489, 2000.

[124] Y. Ning and H. Liu. A general theory of hypothesis tests and confidence regions for sparse high dimensional models. *Annals of Statistics*, 45:158–195, 2017.

[125] S. Oba, M. Sato, I. Takemasa, M. Monden, K.I. Matsubara, and S. Ishii. A Bayesian missing value estimation method for gene expression profile data. *Bioinformatics*, 19(16):2088–2096, 2003.

[126] A. Owen. Karl pearson's meta analysis revisited. *Quality Engineering*, 55:493–494, 2009.

[127] G.P. Patil and S.W. Joshi. *A Dictionary and Bibliography of Discrete Distributions*. Hafner Publishing Co., New York, 1968.

[128] J. Pearl. Bayesian networks: A model of self-activated memory for evidential reasoning. In *Proceedings of the 7th Conference of the Cognitive Science Society, University of California*, Ivine, CA, pp. 329–334, 1985.

[129] J. Pearl. *Probabilistic Reasoning in Intelligent Systems:Networks of Plausible Inference*. Morgan Kaufmann Publishers, Inc., San Francisco, CA, 1988.

[130] J.-P. Pellet and A. Elisseeff. Using Markov blankets for causal structure learning. *Journal of Machine Learning Research*, 9:1295–1342, 2008.

[131] C. Peterson, F. Stingo, and M. Vannucci. Bayesian inference of multiple Gaussian graphical models. *Journal of the American Statistical Association*, 110:159–174, 2015.

[132] W. Poole, D.L. Gibbs, I. Shmulevich, B. Bernard, and T.A. Knijnenburg. Combining dependent p-values with an empirical adaptation of brown's method. *Bioinformatics*, 32(17):i430–i436, 2016.

[133] S. Portnoy. Asymptotic behavior of likelihood methods for exponential families when the number of parameters tends to infinity. *Annals of Statistics*, 16:356–366, 1988.

[134] H. Qiu, F. Han, H. Liu, and B. Caffo. Joint estimation of multiple graphical models from high dimensional time series. *Journal of the Royal Statistical Society. Series B, Statistical Methodology*, 78 2:487–504, 2016.

[135] P. Ravikumar, M. Wainwright, and J. Lafferty. High-dimensional ising model selection using l_1-regularized logistic regression. *Annals of Statistics*, 38:1287–1319, 2009.

[136] A. Rinaldo, L.A. Wasserman, M. G'sell, and J. Lei. Bootstrapping and sample splitting for high-dimensional, assumption-lean inference. *The Annals of Statistics*, 47(6):3438–3469, 2019.

[137] H. Robbins and S. Monro. A stochastic approximation method. *The Annals of Mathematical Statistics*, 22(3):400–407, 1951.

[138] M. Robinson and A. Oshlack. A scaling normalization method for differential expression analysis of RNA-seq data. *Genome Biology*, 11:R25–R25, 2009.

[139] V. Rocková and E.I. George. The spike-and-slab Lasso. *Journal of the American Statistical Association*, 113:431–444, 2018.

[140] L. Ruan, M. Yuan, and H. Zou. Regularized parameter estimation in high-dimensional Gaussian mixture models. *Neural Computation*, 23:1605–1622, 2011.

[141] D.F. Saldana and Y. Feng. SIS: An R package for sure independence screening in ultrahigh-dimensional statistical models. *Journal of Statistical Software*, 83:1–25, 2018.

[142] M. Scutari and J.-B. Denis. *Bayesian Networks with Examples in R*. CRC Press, New York, 2015.

[143] M. Scutari. Learning Bayesian networks with the Bnlearn R package. *Journal of Statistical Software*, 35:1–22, 2009.

[144] R. Sen, A.T. Suresh, K. Shanmugam, A.G. Dimakis, and S. Shakkottai. Model-powered conditional independence test. In: I. Guyon, U. von Luxburg, S. Bengio, H.M. Wallach, R. Fergus, S.V.N. Vishwanathan, and R. Garnett, editors, *Advances in Neural Information Processing Systems 30: Annual Conference on Neural Information Processing Systems 2017*, Long Beach, CA, December 4–9, 2017, pp. 2951–2961, 2017.

[145] R.D. Shah and J. Peters. The hardness of conditional independence testing and the generalised covariance measure. *The Annals of Statistics*, 48(3):1514–1538, 2020.

[146] C. Shi, R. Song, W. Lu, and R. Li. Statistical inference for high-dimensional models via recursive online-score estimation. *Journal of the American Statistical Association*, 116:1307–1318, 2021.

[147] S. Shimizu, P.O. Hoyer, A. Hyvärinen, and A.J. Kerminen. A linear non-Gaussian acyclic model for causal discovery. *Journal of Machine Learning Research*, 7:2003–2030, 2006.

[148] L. Silwal-Pandit, H.K. Vollan, S.F. Chin, O.M. Rueda, S. McKinney, T. Osako, D.A. Quigley, V.N. Kristensen, S. Aparicio, A.L. Borresen-Dale, C. Caldas, and A. Langerod. TP53 mutation spectrum in breast cancer is subtype specific and has distinct prognostic relevance. *Clinical Cancer Research*, 20(13):3569–3580, 2014.

[149] Q. Song and F. Liang. High dimensional variable selection with reciprocal L1-regularization. *Journal of the American Statistical Association*, 110:1607–1620, 2015.

[150] Q. Song and F. Liang. A split-and-merge Bayesian variable selection approach for ultra-high dimensional regression. *Journal of the Royal Statistical Society, Series B*, 77:947–972, 2015.

[151] Q. Song and F. Liang. Nearly optimal Bayesian shrinkage for high-dimensional regression. *Science China Mathematics*, 66(2):409–442, 2023.

[152] P. Spirtes, C. Glymour, and R. Scheines. *Causation, Prediction, and Search*. Springer-Verlag, Berlin, Germany, 1993.

[153] P. Spirtes, C. Glymour, and R. Scheines. *Causation, Prediction, and Search*, (2nd edition). MIT Press, Cambridge, MA, 2000.

[154] W. Stacklies, H. Redestig, M. Scholz, D. Walther, and J. Selbig. pcaMethods: A bioconductor package providing PCA methods for incomplete data. *Bioinformatics*, 23(9):1164–1167, 2007.

[155] N. Städler and P. Bühlmann. Missing values: Sparse inverse covariance estimation and an extension to sparse regression. *Statistics and Computing*, 22:219–235, 2012.

[156] J.D. Storey. A direct approach to false discovery rates. *Journal of the Royal Statistical Society: Series B (Statistical Methodology)*, 64:479–498, 2002.

[157] S. Stouffer, E. Suchman, L.C. DeVinney, S.A. Star, and R.M. Williams. *The American Soldier: Adjustment During Army Life*, Vol. 1. Princeton University Press, Princeton, NJ, 1949.

[158] E.V. Strobl, K. Zhang, and S. Visweswaran. Approximate Kernel-based conditional independence tests for fast non-parametric causal discovery. *Journal of Causal Inference*, 7(1):20180017, 2019.

[159] M. Sultan, M.H. Schulz, H. Richard, A. Magen, A. Klingenhoff, M. Scherf, M. Seifert, T. Borodina, A. Soldatov, D. Parkhomchuk, D. Schmidt, S. O'Keeffe, S. Haas, M. Vingron, H. Lehrach, and M. Yaspo. A global view of gene activity and alternative splicing by deep sequencing of the human transcriptome. *Science*, 321:956–960, 2008.

[160] L. Sun and F. Liang. Markov neighborhood regression for statistical inference of high-dimensional generalized linear models. *Statistics in Medicine*, 41:4057–4078, 2022.

[161] L. Sun, A. Zhang, and F. Liang. Consistent dynamic Bayesian network learning for a fMRI study of emotion processing. *Manuscript*, 2022.

[162] Y. Sun, Q. Song, and F. Liang. Consistent sparse deep learning: Theory and computation. *Journal of the American Statistical Association*, 117(540):1981–1995, 2022.

[163] P. Sur, Y. Chen, and E.J. Candès. The likelihood ratio test in high-dimensional logistic regression is asymptotically a rescaled chi-square. *Probability Theory and Related Fields*, 175:487–558, 2019.

[164] R. Tibshirani. Regression shrinkage and selection via the Lasso. *Journal of the Royal Statistical Society, Series B*, 58:267–288, 1996.

[165] R.J. Tibshirani, A. Rinaldo, R. Tibshirani, and L.A. Wasserman. Uniform asymptotic inference and the bootstrap after model selection. *The Annals of Statistics*, 46(3):1255–1287, 2018.

[166] O.G. Troyanskaya, M.N. Cantor, G. Sherlock, P.O. Brown, T.J. Hastie, R. Tibshirani, D. Botstein, and R.B. Altman. Missing value estimation methods for DNA microarrays. *Bioinformatics*, 17(6):520–525, 2001.

[167] I. Tsamardinos, L.E. Brown, and C.F. Aliferis. The max-min hill-climbing Bayesian network structure learning algorithm. *Machine Learning*, 65(1):31–78, 2006.

[168] I. Tsamardinos, C.F. Aliferis, and A.R. Statnikov. Algorithms for large scale Markov blanket discovery. In: I. Russell and S.M. Haller, editors, *Proceedings of the Sixteenth International Florida Artificial Intelligence Research Society Conference*, St. Augustine, FL, May 12–14, 2003, pp. 376–381. AAAI Press, 2003.

[169] P. Tseng. Convergence of a block coordinate descent method for nondifferentiable minimization. *Journal of Optimization Theory and Applications*, 109:475–494, 2001.

[170] A.B. Tsybakov. *Introduction to Nonparametric Estimation*. Springer, New York, 2009.

[171] S. Uda and H. Kubota. Sparse Gaussian graphical model with missing values. In: *21st Conference of Open Innovations Association, FRUCT 2017*, Helsinki, Finland, November 6–10, 2017, pp. 336–343. IEEE, 2017.

[172] S. van Buuren and K.G.M. Groothuis-Oudshoorn. MICE: Multivariate imputation by chained equations in R. *Journal of Statistical Software*, 45:1–67, 2011.

[173] S. van de Geer, P. Bühlmann, Y. Ritov, and R. Dezeure. On asymptotically optimal confidence regions and tests for high-dimensional models. *Annals of Statistics*, 42(3):1166–1202, 2014.

[174] S. van de Geer, P. Bühlmann, and S. Zhou. The adaptive and the thresholded Lasso for potentially misspecified models (and a lower bound for the Lasso). *Electronic Journal of Statistics*, 5:688–749, 2011.

[175] L. Vandenberghe, S.P. Boyd, and S.-P. Wu. Determinant maximization with linear matrix inequality constraints. *SIAM Journal on Matrix Analysis and Applications*, 19:499–533, 1998.

[176] T. Verma and J. Pearl. Equivalence and synthesis of causal models. *Uncertainty in Artificial Intelligence*, 6:255–268, 1991.

[177] T. Verma and J. Pearl. An algorithm for deciding if a set of observed independencies has a causal explanation. In: D. Dubois and M.P. Wellman, editors, *UAI '92: Proceedings of the Eighth Annual Conference on Uncertainty in Artificial Intelligence, Stanford University*, Stanford, CA, July 17–19, 1992, pp. 323–330. Morgan Kaufmann, 1992.

[178] Y.-W. Wan, G.I. Allen, Y. Baker, E. Yang, P. Ravikumar, and Z. Liu. XMRF: An R package to fit Markov networks to high-throughput genetics data. *BMC Systems Biology*, 10(Suppl 3):69, 2015.

[179] M. West, C. Blanchette, H. Dressman, E. Huang, S. Ishida, R. Spang, H. Zuzan, J. Olson, J. Marks, and J. Nevins. Predicting the clinical status of human breast cancer by using gene expression profiles. *Proceedings of the National Academy of Sciences of the United States of America*, 98:1146 – 11467, 2001.

[180] D.J. Wilson. The harmonic mean p-value for combining dependent tests. *Proceedings of the National Academy of Sciences of the United States of America*, 116:1195–1200, 2019.

[181] D. Witten, J. Friedman, and N. Simon. New insights and faster computations for the graphical Lasso. *Journal of Computational and Graphical Statistics*, 20:892–900, 2011.

[182] Y. Xie, Y. Liu, and W. Valdar. Joint estimation of multiple dependent Gaussian graphical models with applications to mouse genomics. *Biometrika*, 103:493–511, 2016.

[183] S. Xu, B. Jia, and F. Liang. Learning moral graphs in construction of high-dimensional Bayesian networks for mixed data. *Neural Computation*, 31:1183–1214, 2019.

[184] J. Xue and F. Liang. A robust model-free feature screening method for ultrahigh-dimensional data. *Journal of Computational and Graphical Statistics*, 26:803–813, 2017.

[185] E. Yang, G. Allen, Z. Liu, and P.K. Ravikumar. Graphical models via generalized linear models. In: S.A. Solla, K.-R. Müller, and T.K. Leenpp, editors, *Advances in Neural Information Processing Systems*, pp. 1358–1366. MIT Press, Cambridge, MA, 2012.

[186] E. Yang, P. Ravikumar, G.I. Allen, Y. Baker, Y.-W. Wan, and Z. Liu. A general framework for mixed graphical models. arXiv:1411.0288v1, 2014.

[187] E. Yang, P. Ravikumar, G.I. Allen, and Z. Liu. On poisson graphical models. In: C.J.C. Burges, L. Bottou, Z. Ghahramani, and K.Q. Weinberger, editors, *Advances in Neural Information Processing Systems 26: 27th Annual Conference on Neural Information Processing Systems 2013*, Lake Tahoe, NV, December 5–8, 2013, pp. 1718–1726, 2013.

[188] T.-H. Yang, C.-C. Wang, Y.-C. Wang, and W.-S. Wu. YTRP: A repository for yeast transcriptional regulatory pathways. *Database: The Journal of Biological Databases and Curation*, 2014:bau014, 2014.

[189] S. Yaramakala and D. Margaritis. Speculative Markov blanket discovery for optimal feature selection. In *Proceedings of the 5th IEEE International Conference on Data Mining (ICDM 2005)*, Houston, TX, November 27–30, 2005, pp. 809–812. IEEE Computer Society, 2005.

[190] J. Yin and H. Li. A sparse conditional Gaussian graphical model for analysis of genetical genomics data. *The Annals of Applied Statistics*, 5(4):2630–2650, 2011.

[191] Y. Yu, J. Chen, T. Gao, and M. Yu. DAG-GNN: DAG structure learning with graph neural networks. In: K. Chaudhuri and R. Salakhutdinov, editors, *Proceedings of the 36th International Conference on Machine Learning, ICML 2019*, Long Beach, CA, June 9–15, 2019, volume 97 of *Proceedings of Machine Learning Research*, pp. 7154–7163. PMLR, 2019.

[192] M. Yuan and Y. Lin. Model selection and estimation in regression with grouped variables. *Journal of the Royal Statistical Society: Series B (Statistical Methodology)*, 68:49–67, 2006.

[193] M. Yuan and Y. Lin. Model selection and estimation in the Gaussian graphical model. *Biometrika*, 94(1):19–35, 2007.

[194] D. Zaykin. Optimally weighted z-test is a powerful method for combining probabilities in meta-analysis. *Journal of Evolutionary Biology*, 24(8):1836–1841, 2011.

[195] C.-H. Zhang. Nearly unbiased variable selection under minimax concave penalty. *Annals of Statistics*, 38:894–942, 2010.

[196] C.-H. Zhang and S.S. Zhang. Confidence intervals for low dimensional parameters in high dimensional linear models. *Journal of the Royal Statistical Society: Series B (Statistical Methodology)*, 76(1):217–242, 2014.

[197] K. Zhang, J. Peters, D. Janzing, and B. Schölkopf. Kernel-based conditional independence test and application in causal discovery. In: F.G. Cozman and A. Pfeffer, editors, *UAI 2011, Proceedings of the Twenty-Seventh Conference on Uncertainty in Artificial Intelligence*, Barcelona, Spain, July 14–17, 2011, pp. 804–813. AUAI Press, 2011.

[198] Q. Zhang, S. Filippi, S.R. Flaxman, and D. Sejdinovic. Feature-to-feature regression for a two-step conditional independence test. In: G. Elidan, K. Kersting, and A. Ihler, editors, *Proceedings of the Thirty-Third Conference on Uncertainty in Artificial Intelligence, UAI 2017*, Sydney, Australia, August 11–15, 2017. AUAI Press, 2017.

[199] P. Zhao and B. Yu. On model selection consistency of Lasso. *Journal of Machine Learning Research*, 7:2541–2563, 2006.

[200] T. Zhao, H. Liu, K. Roeder, J. Lafferty, and L. Wasserman. The huge package for high-dimensional undirected graph estimation in R. *Journal of Machine Learning Research (JMLR)*, 13:1059–1062, 2012.

[201] X. Zheng, B. Aragam, P. Ravikumar, and E.P. Xing. Dags with NO TEARS: Continuous optimization for structure learning. In: S. Bengio, H.M. Wallach, H. Larochelle, K. Grauman, N. Cesa-Bianchi, and R. Garnett, editors, *Advances in Neural Information Processing Systems 31: Annual Conference on Neural Information Processing Systems 2018, NeurIPS 2018*, Montréal, Canada, December 3–8, 2018, pp. 9492–9503, 2018.

[202] X. Zheng, C. Dan, B. Aragam, P. Ravikumar, and E.P. Xing. Learning sparse nonparametric dags. In: S. Chiappa and R. Calandra, editors, *The 23rd International Conference on Artificial Intelligence and Statistics, AISTATS 2020*, Palermo, Sicily, Italy, August 26–28, 2020, volume 108 of *Proceedings of Machine Learning Research*, pp. 3414–3425. PMLR, 2020.

[203] S. Zhou, J. Lafferty, and L. Wasserman. Time varying undirected graphs. *Machine Learning*, 80:295–319, 2010.

[204] G. Zoppoli, M. Regairaz, E. Leo, W.C. Reinhold, S. Varma, A. Ballestrero, J.H. Doroshow, and Y. Pommier. Putative DNA/RNA helicase schlafen11 (SLFN11) sensitizes cancer cells to DNA-damaging agents. *Proceedings of the National Academy of Sciences USA*, 109:15030–15035, 2012.

[205] H. Zou. The adaptive Lasso and its oracle properties. *Journal of the American Statistical Association*, 101:1418–1429, 2006.

[206] H. Zou and T.J. Hastie. Regularization and variable selection via the elastic net. *Journal of the Royal Statistical Society: Series B (Statistical Methodology)*, 67:301–320, 2005.

Index

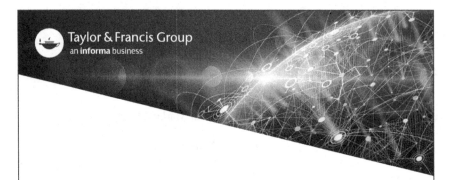